SpringerBriefs in Applied Sciences and Technology

Series editor

Janusz Kacprzyk, Polish Academy of Sciences, Systems Research Institute, Warsaw, Poland

SpringerBriefs present concise summaries of cutting-edge research and practical applications across a wide spectrum of fields. Featuring compact volumes of 50–125 pages, the series covers a range of content from professional to academic. Typical publications can be:

- A timely report of state-of-the art methods
- An introduction to or a manual for the application of mathematical or computer techniques
- A bridge between new research results, as published in journal articles
- A snapshot of a hot or emerging topic
- An in-depth case study
- A presentation of core concepts that students must understand in order to make independent contributions

SpringerBriefs are characterized by fast, global electronic dissemination, standard publishing contracts, standardized manuscript preparation and formatting guidelines, and expedited production schedules.

On the one hand, **SpringerBriefs in Applied Sciences and Technology** are devoted to the publication of fundamentals and applications within the different classical engineering disciplines as well as in interdisciplinary fields that recently emerged between these areas. On the other hand, as the boundary separating fundamental research and applied technology is more and more dissolving, this series is particularly open to trans-disciplinary topics between fundamental science and engineering.

Indexed by EI-Compendex and Springerlink.

More information about this series at http://www.springer.com/series/8884

Brajesh Kumar Kaushik · Shivam Verma
Anant Aravind Kulkarni · Sanjay Prajapati

Next Generation Spin Torque Memories

Springer

Brajesh Kumar Kaushik
Department of Electronics
  and Communication Engineering
Indian Institute of Technology Roorkee
Roorkee, Uttarakhand
India

Anant Aravind Kulkarni
Department of Electronics
  and Communication Engineering
Indian Institute of Technology Roorkee
Roorkee, Uttarakhand
India

Shivam Verma
Department of Electronics
  and Communication Engineering
Indian Institute of Technology Roorkee
Roorkee, Uttarakhand
India

Sanjay Prajapati
Department of Electronics
  and Communication Engineering
Indian Institute of Technology Roorkee
Roorkee, Uttarakhand
India

ISSN 2191-530X   ISSN 2191-5318   (electronic)
SpringerBriefs in Applied Sciences and Technology
ISBN 978-981-10-2719-2   ISBN 978-981-10-2720-8   (eBook)
DOI 10.1007/978-981-10-2720-8

Library of Congress Control Number: 2016963429

Printed on acid-free paper

This Springer imprint is published by Springer Nature
The registered company is Springer Nature Singapore Pte Ltd.
The registered company address is: 152 Beach Road, #21-01/04 Gateway East, Singapore 189721, Singapore

To my Father Late Mr. Jai Prakash Kaushik for his affection and untiring efforts in my upbringing. Dear father, I miss you a lot.

Brajesh Kumar Kaushik

To my mother for her selfless love and support.

Shivam Verma

To my Uncle Late Dr. Vasant Deshpande.

Anant Aravind Kulkarni

To my family members for their understanding, continuous motivation, and unwavering support that allowed me to focus and complete this work successfully.

Sanjay Prajapati

Preface

Memory technology has gained a pivotal position in the contemporary semiconductor computation industry. The last few decades have witnessed remarkable changes in the role and importance of various semiconductor memory technologies based on key parameters such as speed, power, density, and cost per bit of storage devices. The punch cards used in the beginning of the last century required an access time of nearly one second for the card of 1 KB memory size; on the contrary, modern finger-sized flash memories are able to achieve a speed of nearly Gbps. At present, modern computing systems employ magnetic memories such as hard disk drives (HDDs) and semiconductor memories such as static random access memory (SRAM), dynamic random access memory (DRAM), and flash memories at different levels of memory hierarchy.

In the diminishing era of Moore's law on the silicon road map, all the aforementioned memory technologies are still unable to cope with the speed of contemporary processing devices. Nonvolatile HDDs with high storage capacity are bulkier and sluggish; hence, they remain at the bottom of the memory hierarchy. At the top of the memory hierarchy, SRAM exhibits highest speed near to 1 ns; however, SRAM lacks in storage capacity, dissipates very high standby leakage power, and is volatile in nature. DRAM faces the problems of increasing refresh current and complex physical fabrication process. Flash memories suffer from excess write power, sluggish write speed, reliability and inadequate endurance issues. Therefore, it is imperative for the researchers to develop alternative nonvolatile memory technology solutions to meet the requirements of futuristic high-speed communication and computational applications. Emerging nonvolatile memory (NVM) technologies, including spin-torque magnetic RAM (ST-MRAM), ferroelectric RAM (FeRAM), phase change RAM (PCRAM), and resistive RAM (RRAM), are some of the competing and promising contenders as memory technologies for futuristic high-speed high-density on-chip storage applications.

Among all the aforementioned emerging NVM technologies, spintronic memory technologies have become the center of research attraction due to their possession of all the features of a universal memory such as nonvolatility, higher densities, enhanced performance, low power dissipation, unlimited endurance, high retention, and CMOS-compatible fabrication process. This book describes all these facets of various spintronic-based magnetic memories such as spin transfer torque (STT), spin orbit torque (SOT), domain wall (DW) MRAMs, and sequential racetrack memories (RM). A spintronic device known as magnetic tunnel junction (MTJ) which utilizes the spin of electrons as a state variable is the key element in all kinds of emerging spintronic MRAMs. Specifically, perpendicular magnetic anisotropy (PMA)-based MTJ devices with their scalable architectures in the nanoscale regime and low-power magnetic switching requirements have gained stupendous interest among the research community. All these spintronic memories require an MTJ with one or two access transistors; hence, a huge footprint area saving can be achieved in comparison with conventional semiconductor memories.

In the initial phase of evolution, STT-MRAMs have encountered the problems of high switching threshold current and larger access device dimensions; however, with the advent of PMA devices, recently, the switching current requirements have been made low enough to reach at the level of 10 μA with access time near to the SRAMs which is best suitable for on-chip embedded memory applications. Significant efforts are being made to decrease the access device dimensions using metal oxide semiconductor (MOS) technologies other than the planar architectures with larger driving capabilities. This book describes the gate-all-around (GAA) MOSFET (MOS field effect transistor), a vertical 3-dimensional (3D) nanowire access device, employed to drive an MTJ element resulting an STT-MRAM with minimum footprint area of $4F^2$ (F is the feature size). SOT-MRAM provides the energy-efficient memory technology solution with different read and write path optimization. Multilevel-cell (MLC) MRAMs offer high density at reduced cost per bit with the help of series or parallel arrangements of the storage elements, i.e. MTJs. Racetrack memory utilizes the concept of magnetic domain wall motion within a nanowire and has the potential to cover the entire memory hierarchy. Racetrack memories (RM) possess all the features of a universal memory architecture, e.g., high density of HDD, high speed as SRAMs, unlimited endurance, high retention period, nonvolatility, lower switching and operation power, and above all capability of 3D integration with the complementary metal oxide semiconductor (CMOS)-compatible fabrication process. Therefore, RM can become a revolutionary memory technology which is best suitable for high-speed embedded communication and computing devices.

This book comprehensively describes all the aspects of the spintronic memories and discusses the importance of various key physical and electrical parameters affecting the performance of the memories to be considered while designing and optimizations. Chapter 1 provides a brief introduction to all the emerging spintronic memories covered in this book. Chapter 2 provides the detailed insight into next-generation 3D vertical silicon nanowire (NW)-based STT-MRAMs with vertical GAA select device. With the help of a case study, a performance comparison between GAA device based and planar MOSFET-based STT-MRAMs has been demonstrated. The evident advantages of GAA-based vertical silicon nanowire STT-MRAMs in terms of write margins, power dissipation, and 2D array density of $4F^2$ have been presented in the chapter. Chapter 3 provides a comprehensive description of SOT-MRAM including the understanding of basic operational mechanisms and modeling of the same. A comparative performance analysis of STT- and SOT-MRAMs has been depicted at the end of the chapter. Chapter 4 demonstrates series and parallel configurations of the MLC MRAMs using in-plane magnetic anisotropy (IMA) and PMA MTJ devices. This chapter provides a detailed comparison of all the possible combinations of MLC structures with the help of simulations carried out using the SPICE-compatible simulation framework. Chapter 5 reveals the importance of most attractive and emerging racetrack memory technology which has the potential to revolutionize the computing industry. A fundamental concept of domain wall motion which is playing a key role in the racetrack memory is discussed broadly in the chapter. At the end the chapter, the important features of racetrack memory to realize the concept of "logic-in-memory" are discussed.

We acknowledge the support of all the faculty members and students of Department of Electronics and Communication Engineering, Indian Institute of Technology Roorkee, for their inputs and feedbacks in accomplishing this book to utmost satisfaction and all possible expectations of a reader.

Roorkee, India   Brajesh Kumar Kaushik
                                                                              Shivam Verma
                                                                      Anant Aravind Kulkarni
                                                                            Sanjay Prajapati

Contents

About the Authors

Brajesh Kumar Kaushik received his Bachelor of Engineering degree in Electronics and Communication Engineering from D.C.R. University of Science and Technology, (formerly *C. R. State College of Engineering*) Murthal, Haryana, in 1994. He received Master of Technology degree in Engineering Systems, from Dayalbagh Educational Institute, Agra, in 1997, and Doctorate of Philosophy (Ph.D.) degree in 2007 under AICTE-QIP scheme from Indian Institute of Technology Roorkee, India. He served Vinytics Peripherals Pvt. Ltd., Delhi, as Research and Development Engineer in microprocessor, microcontroller, and DSP processor-based system design. He joined Department of Electronics and Communication Engineering, G.B. Pant Engineering College, Pauri Garhwal, Uttarakhand, India, as Lecturer in July, 1998, where later, he served as Assistant Professor from May, 2005 to May, 2006 and Associate Professor from May, 2006 to December, 2009. He joined Department of Electronics and Communication Engineering, Indian Institute of Technology Roorkee as Assistant Professor in December, 2009; where since April, 2014, he is working as Associate Professor. He has extensively published in several national and international journals and conferences of repute. He has also authored/co-authored several books and book chapters. He is reviewer of many international journals belonging to various publications such as IEEE, IET, Elsevier, Springer, Taylor and Francis, Emerald, ETRI, and PIER. He has also served as General Chair, Technical Chair, and Keynote Speaker in many reputed international and national conferences. Dr. Kaushik is *Senior Member* of IEEE and member of many expert committees constituted by government and non-government organizations. He holds the position of Editor and Editor-in-Chief of various journals in the field of VLSI and microelectronics such as *International Journal of VLSI Design & Communication Systems (VLSICS)*, AIRCC Publishing Corporation. He also holds the position of Editor of *Microelectronics Journal (MEJ)*, Elsevier Inc.; *Journal of Engineering, Design and Technology (JEDT)*, Emerald Group Publishing Limited; *Journal of Electrical and Electronics Engineering Research (JEEER)*; and Academic Journals. He has received many awards and recognitions from the International

Biographical Center (IBC), Cambridge. His name has been listed in Marquis Who's Who in Science and Engineering® and Marquis Who's Who in the World®. Dr. Kaushik has been conferred with Distinguished Lecturer award of IEEE Electron Devices Society (EDS) to offer EDS Chapters with quality lectures in his research domain. His research interests are in the areas of high-speed interconnects, low-power VLSI design, memory design, carbon nanotube-based designs, organic electronics, FinFET device circuit co-design, electronic design automation (EDA), spintronics-based devices, circuits and computing, image processing, and optics and photonics based devices.

Shivam Verma received his Bachelor of Engineering degree in Electronics and Communication Engineering from Shri Vaishnav Institute of Technology and Science, Indore, India, and the Master of Technology degree in Microelectronics from Indian Institute of Technology Varanasi, India, in 2010 and 2012, respectively. He received Doctorate of Philosophy (Ph.D.) degree from Indian Institute of Technology Roorkee, India, in 2016. He has published several papers in IEEE journals. His current research interests include spin-transfer torque-based devices and all spin logic.

Anant Aravind Kulkarni received his Bachelor of Engineering degree in Electronics Engineering from Shri Guru Gobind Singhji College of Engineering and Technology, Nanded, Maharashtra, India, in 2002; his first Master of Technology degree in Electrical Engineering from Uttar Pradesh Technical University (presently Dr. A.P.J. Abdul Kalam Technical University), Lucknow, India, in 2009; and his second Master of Technology degree in Microelectronics and Very-large-scale Integration Design from the Technocrat Institute of Technology, Bhopal, India, in 2013. He is presently pursuing Doctorate of Philosophy (Ph.D.) degree from Indian Institute of Technology Roorkee, India. He joined Department of Electronics and Communication Engineering, Marathwada Institute of Technology, Bulandshahr, India, as Lecturer in July, 2003 and served there till January, 2007. He worked in Electronics and Communication Engineering Department, Meerut Institute of Engineering and Technology, Meerut, India, as Lecturer from February, 2007 to January, 2008. He joined Electrical, Electronics, and Power Engineering Department, College of Engineering, Ambajogai, India as Senior Lecturer in February, 2008, where since June, 2009, he is working as Assistant Professor. His current research interests include spintronics based devices, circuits, and computing.

Sanjay Prajapati received his Bachelor of Engineering degree in Electronics and Communication Engineering from Government Engineering College, Modasa, Gujarat, India, in 1996 and Master of Technology degree in Very-large-scale Integration Design from Nirma University, Ahmedabad, Gujarat, India, in 2010. He is pursuing Doctorate of Philosophy (Ph.D.) degree from Indian Institute of Technology Roorkee, India, since July 2015. He served as a Lecturer in the Department of Electronics and Communication Engineering, Government Polytechnic, Surat, India, from October, 1998 to October, 2004. He joined as Assistant Professor at

Vishwakarma Government Engineering College, Ahmedabad, India, in October, 2004 and later promoted as Associate Professor in September, 2012, at Government Engineering College, Dahod, affiliated to Gujarat Technological University, Ahmedabad, India. He has attended several workshops, seminars, and faculty development programs of national level. He has more than 18 years of academic and research experience. His current research interest includes spintronic device modeling, memory, and logic design.

Chapter 1
Emerging Memory Technologies

1.1 Introduction

In conventional memory hierarchy, memories near and away from the processor provide short and long latencies (see Fig. 1.1), respectively. Static random access memory (SRAM), dynamic random access memory (DRAM), and hard disks have provided the foundation for the cache, main memory, and external storage, respectively. DRAM [1] is a high density volatile memory due to its simple bit-cell configuration made up of single transistor in series with a single capacitor, and it needs continuous source of energy to retain the information. However, DRAM has very high power dissipation due to off-state leakage current. The other memory technology is SRAM (static RAM) [2], which stores one bit of information in a flip-flop. SRAM does not require periodic refresh to retain the data. In comparison to DRAM, the SRAM has low operational power dissipation; operates at high speed; and it is very reliable. However, SRAM cell needs six transistors to store one bit of information, which increases the cell cost and lowers the overall density. Moreover, SRAM has high leakage power dissipation. The hard disk memories have high access latency and power dissipation due to off-chip memory access.

In recent years, computing power is increased significantly due to the use of multicore processors [3]. However, high memory capacity is the stringent requirement in many complex applications, which puts restrictions on the existing memory systems. In addition, parallel execution of multiple applications needs a high performance memory system. Therefore, to get rid of the aforementioned obstacles, the memory hierarchy is required to be re-designed to enhance the overall performance of the computing systems. The flash memories [4] such as NAND and NOR have shown good prospects over SRAM and DRAM due to their nonvolatile nature. However, NOR has poor write and erase time, and NAND has non-standard interface and complicated management. Therefore, researchers are intensively working on the emerging non-volatile memories [5] to provide the novel possibilities, which could replace conventional non-volatile memories in near future.

© The Author(s) 2017
B.K. Kaushik et al., *Next Generation Spin Torque Memories*,
SpringerBriefs in Applied Sciences and Technology,
DOI 10.1007/978-981-10-2720-8_1

Fig. 1.1 Conventional
memory hierarchy

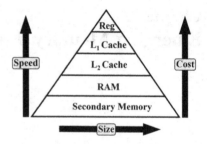

1.2   Non-volatile Memories

The major disadvantage of SRAM and DRAM is that they require continuous source of energy to retain the information due to their volatile nature. Therefore, these power hungry memories create obstacles in achieving ultra-low power computing. The other issue with SRAM is low density due to six MOS (metal oxide semiconductor) transistors required to realize a SRAM cell. Therefore, the high density non-volatile memories become very important to achieve high performance for ultra-low power computing applications. The evolution of emerging non-volatile memories is shown in Fig. 1.2.

Novel non-volatile memories differ from the mature memories in terms of materials and switching mechanism. The novel materials for the non-volatile memories are ferroelectric dielectrics, ferroelectric metals, chalcogenides, transitional metal oxides, carbon materials, etc. The switching mechanisms include quantum mechanical phenomenon, ionic reactions, phase transition, molecular reconfiguration, etc. Unlike SRAM, most of the non-volatile memories are based on the two terminal switching element. The two-terminal switching element is suitable

Fig. 1.2 RAM classification

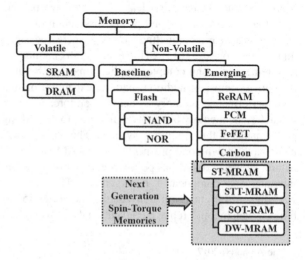

for the high density memory architectures. In the subsequent subsections, the emerging non-volatile memories are briefly presented.

1.2.1   Phase Change Memory

The PCM (phase change memory) [6] structure is shown in Fig. 1.3. It consists of a non-volatile storage element and an access device. The storage element is made up of chalcogenide material $Ge_2Sb_2Te_5$ (GST), thermal insulator, and heater. To store 1-bit information, chalcogenide material is used. Phase change materials show the phase transition characteristics at smaller dimensions, higher crystalline temperature, lower thermal conductivity, and improved endurance. Moreover, the materials used for the phase change memory are scalable with reduced thermal conductivity. PCM cells need to be designed to simplify processing, optimize power efficiency, and reduce reset current. The access device as shown in Fig. 1.3 is utilized to read the state of the phase change material. The use of vertical access device reduces the size of PCM cell. Therefore, the PCM performance depends on the property of phase change material and memory cell design.

The switching power of PCM can be minimized by contact or volume minimization approach. The switching speed and endurance of the PCM are ~ 100 ns and 10^9 cycles, respectively. However, the large switching current requirement prevents the smaller size of the PCM bit cell. Moreover, the cost of the PCM is very high as compared to the other incumbent memories. The applications of PCM memory are storage class memory, ternary content memory, neuromorphic computing etc.

Fig. 1.3 PCM bit cell architecture

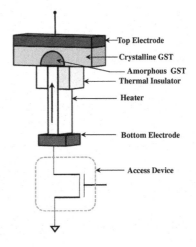

1.2.2 Resistive RAM

The materials such as chalcogenide, semiconductors, oxides nitrides, and organic materials have shown the properties of resistive memory through the distinct resistance states. The bit cell architecture of ReRAM (resistive RAM) [7] is shown in Fig. 1.4. The two terminal storage device consists of a metal oxide sandwiched between two electrodes. The metal oxides could be NiO_x, HfO_x, TiO_x, TaO_x, or $Pr_xCa_{1-x}MnO_3$. The ReRAM works on the principle of dielectric material conductivity. A dielectric becomes conductive through filament formation when a current is passed at a sufficiently high voltage. The broken and re-formed filament represents high and low resistance, respectively. The conduction occurs due to the different mechanisms such as vacancy [8] or metal defect migration [9]. The filament formation and destruction represent the two distinct states of bit information. The configuration of filament formed in the metal oxides varies from cycle to cycle and cell to cell in terms of position, dimension, and configuration of the filament. The filament shows stochastic switching behavior due to variation in the metal oxide properties. The stochastic behavior of ReRAM produces large variation in switching voltages. In addition, the balance is required between important design parameters such as speed and power. Moreover, ReRAM is facing the challenge of reliability, variability, and failure mechanisms [10].

Fig. 1.4 ReRAM bit cell architecture

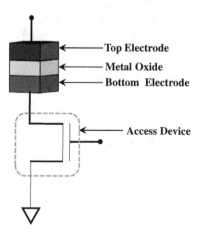

1.2.3 Ferroelectric RAM

An FeRAM (ferroelectric RAM) structure [11] is shown in Fig. 1.5. The ferroelectric polarization is used to represent the data storage. The channel conductance depends on polarization of the ferroelectric material. The overall switching power of the FeRAM is very low due to the field driven gate control with minimum leakage current. The depolarization field, gate leakage, and inability to scale the ferroelectric dielectrics inhibited the further exploration of FeRAM. However, the doped HfO_x [12] as ferroelectric material has shown tremendous potential to realize FeRAM as one of the emerging next generation memories. The FeRAM memory is fabricated in front-end-of-line (FEOL).

1.2.4 Magnetoresistive RAM

An MRAM (magnetoresistive RAM) [13] consists of a non-magnetic spacer sandwiched between two ferromagnetic layers as shown in Fig. 1.6. The non-magnetic spacer could be a tunnel barrier. Magnetization of one of the ferromagnetic layers is pinned and that of the other free. The magnetic polarization of the free layer can be altered either by external magnetic field or by imparting spin torque to utilize it as bit storage. The overall resistance of the layered structure depends on the parallel and anti-parallel configurations of the layers on the either sides of the nonmagnetic spacer. The resistance is minimum (state '1') and maximum (state '0') for the parallel and anti-parallel configurations, respectively. The bit information is sensed through the access device by passing the electric current from side to side of the layered structure.

The '*bit write*' is one of the bottlenecks in MRAM design [14]. Earlier, the external magnetic fields were used to switch the free layer magnetization. However, free layer switching by external magnetic field has several disadvantages. First, it is very difficult to scale the MRAM due to the external magnetic field. Second,

(a) **(b)**

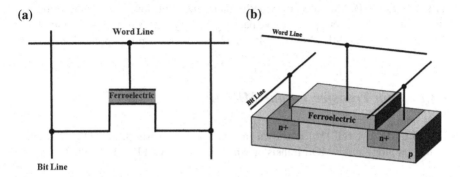

Fig. 1.5 FeRAM bit cell architecture

Fig. 1.6 MRAM bit cell architecture

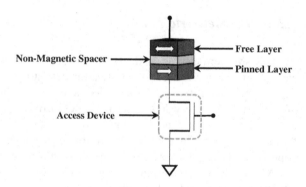

additional power overhead for the magnetic field generation reduces the overall efficiency of the MRAM. To get rid of the obstacles due to the external magnetic field, the ST (spin torque) based MRAMs [15] have emerged as next generation memory technologies to offer scalability, high efficiency, and improved endurance.

This book covers the exploration of next generation ST-MRAMs due to their unique characteristics of scalability, non-volatility, and improved endurance over the other emerging non-volatile memory technologies.

1.3   Spin Torque Based Memories

The ST memories have shown tremendous potential to resolve the issues faced by other emerging memory technologies and have laid the foundation to utilize the concept of material-device-circuit co-design to a greater extent. The ST-based memories are classified as spin transfer torque (STT), spin orbit torque (SOT), and domain wall (DW) memories. Using the nonvolatile features of the MTJ devices with switching ability, STT and SOT-MRAMs have shown the potential to implement the concept of *"on-chip memory-logic"*. The concept of DW memories is based on the magnetic domain wall movement from one magnetic domain to other. STT/SOT/DW-MRAMs are non-volatile memories with low power consumption, high endurance, higher speed, good packing density, and scalability. These characteristics make ST-MRAMs suitable for ultramodern memory applications.

1.3.1   Spin Transfer Torque MRAM

The STT MRAM cell [16] is shown in Fig. 1.7. It consists of an MTJ (magnetic tunnel junction) and a MOS transistor which is termed as 1T-1MTJ cell. The word

Fig. 1.7 STT-MRAM

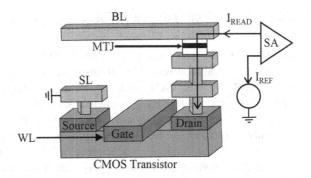

line is connected to the gate of the transistor to access the bit information stored in the MTJ. The bit line (BL) and source line (SL) are connected to the free layer of MTJ and source terminal of the transistor, respectively. The geometry dependent magnetic anisotropy [17] can be classified in to two categories: in-plane magnetic anisotropy (IMA) and perpendicular magnetic anisotropy (PMA). The magnitude of current I_{READ} as shown in Fig. 1.7 depends on the MTJ resistance, which varies with AP and P configurations of the free and pinned layers of the MTJ. Sense amplifier (SA) reads the bit information stored in the free layer of the MTJ.

In STT MRAMs, the cell operation involves bidirectional switching between two distinct resistive states, i.e., R_P and R_{AP}. For AP to P switching, the transistor facilitates a current flow from the BL to SL. Consider the electrons flow from the SL to BL through access transistor and MTJ. The pinned layer acts as filter to injected high density spin electrons through the tunnel barrier. The nonmagnetic tunnel barrier preserves the spin coherence of majority spins to be injected into the free layer of the MTJ. The injected electrons exert STT (spin transfer torque) on the free layer to switch the polarization. The things occur in different manner when current flow is reversed. Electrons passed through the free layer get polarized in the direction of the free layer magnetization. The electrons with spin polarization opposite to the direction of the pinned layer are reflected back and others are passed through the pinned layer. The reflected electrons if large in number, exert a significant torque on the free layer. Therefore, the MTJ shows asymmetric behavior due to variation in magnetization of the free layer. Unlike a conventional switch, MTJ requires different current densities for P to AP and AP to P switching.

Although, the STT-MRAM has emerged as the strongest contender as next generation emerging memory technology, it has several challenges. The STT-MRAM has thinner tunnel barrier, which raises the reliability issues due to the common read and write paths. Therefore, there is possibility of false write operation known as 'read disturb'. Moreover, the separate optimization of the write path is not possible. Therefore, the aforementioned issues create obstacles in making the STT-MRAM a viable alternative for the next generation memory applications such as cache stack and embedded memory architectures.

1.3.2 Spin Orbit Torque MRAM

Recently, a novel spin orbit torque MRAM (SOT-MRAM) [18] architecture, as shown in Fig. 1.8, is presented to get rid-of the problems encountered in the design of STT-MRAM. The SOT structure uses the Rashba and spin-Hall effects (SHE) [19] to flip the free layer magnetization. The SOT devices have the separate read and write current paths to solve the very crucial problem of STT architecture inherently. Moreover, the SOT-MRAM is very much energy efficient and exhibits faster write access due to independent write path optimization. Since, the SOT-MRAM provides the reliable, energy efficient, and fast memory technology solution, it is emerged as a strong contender to replace SRAM in cache memory.

SOT device structure utilizes a Hall metal (HM) attached at the top or bottom of the free layer (FL) of the MTJ as shown in Fig. 1.8. A strong spin-orbit interaction (SOI) can take place at the FL/HM interface when an in-plane current is passed though the HM. This interaction produces an anti-damping torque and a field like torque due to SHE and Rashbha effect, respectively. The strength of the SOI depends on the magnitude of the current passing through the HM. A current with sufficient strength through the HM can switch the magnetization of the FL of the MTJ. Furthermore, the SOI depends on the choice of the material of HM, resistivity of the HM, dimensions of the HM and the MTJ, and area of the FL/HM interface. Since, the hyperfine interaction is proportional to the atomic number of the metal, an HM with higher atomic number (Z) is preferable. The most versatile HM materials used in the SOT-MRAM cell are the tantalum (Ta), tungsten (W), and platinum (Pt).

Fig. 1.8 SOT-MRAM

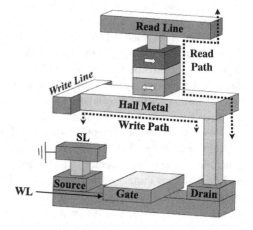

Fig. 1.9 DW-MRAM

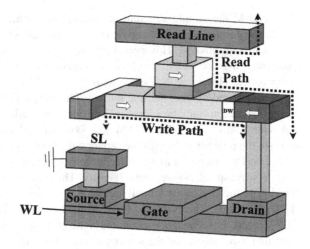

1.3.3 Domain Wall MRAM

The DW-MRAM [20] is shown in Fig. 1.9. The device structure has DW (domain wall) based MTJ with two complementary pinned layers on either sides of the free layer. Basically, a domain wall acts as a boundary between magnetic domains. The energy required for the domain wall is the difference between magnetic moments before and after the domain wall creation. The width of domain wall depends on the anisotropy of the material. The domain wall movement between two pinned layers, as shown in Fig. 1.9, depends on the direction of the current flow between the two pinned layers. If the conventional current flows from the left to right, the magnetization of the free layer sets to the polarization of the right pinned layer and vice versa. The speed of the DW movement is proportional to the magnitude of the current. The read operation is performed by passing the current through the pinned layer of MTJ, tunnel barrier, and DW structure. The read current senses the resistance of the MTJ depending on the relative orientation of the top pinned layer and free layer of the DW structure. The major benefit of DW-MRAM is the low write path resistance, which reduces the write path energy dissipation and improves reliability due to separate read and write paths. Moreover, the width of tunnel barrier can be increased to improve the *TMR* (tunnel magneto-resistance) ratio.

1.4 Comparison of Emerging Memory Technologies

The each one of the emerging nonvolatile memories introduced in this chapter has advantages and challenges. The FeRAM has highest integration density due to 1T memory structure. The major advantages of FeRAM are low-power and high performance storage. The most evolved and mature nonvolatile memory is PCM. Its performance could be improved further by reducing the switching power.

The ReRAM has simple structure, low cost, and higher density. Moreover, it has versatile materials, structures, and behaviors. The ReRAM has better scalability than PCM and STT-MRAM. However, the reliability issues need to be addressed. The most emerging non-volatile memory technology is STT-MRAM due to its better performance. However, the issues with the STT-MRAM are MTJ reliability and BEOL (back-end-of-line) thermal budget. The MTJ reliability issue and high write power dissipation could be addressed by path optimization through the separate read and write paths. The SOT-MRAM and DW-MRAM are the two novel ways to improve the reliability through the separate read-write paths optimization.

The performance of a memory technology is related to the different factors such as scalability, speed, power, and reliability. The scalability of a memory technology depends on the size of the bit-cell architecture, memory architecture, and feasibility of 3D integration. The PCM and ReRAM have better scalability as compared to the STT-MRAM. For a memory technology, the write and read speeds are subjected to switching mechanism and the sensing circuit, respectively. STT-MRAM has better switching mechanism than other memories. On the power front, the PCM, STT-MRAM, and ReRAM have high, moderate, and low write power dissipation, respectively. The STT-MRAM has better retention, endurance, and variability with respect to other memories.

1.5 Chapter Summary

In this chapter, the emerging non-volatile memory technologies are introduced. The bit-cell architectures and operations are presented for ReRAM, PCM, STT-MRAM, SOT-MRAM, and DW-MRAM. These memory technologies provide a wide range of performance, maturity, and scalability. The spin-torque based memories such as STT-MRAM and its improved versions SOT-MRAM and DW-MRAM are emerged as promising contenders for next generation non-volatile memories due to their high density, endurance, and better retention for on chip 'memory-logic' applications. Therefore, in the subsequent chapters the STT-MRAM, SOT-MRAM, MLC-MRAM (multi-level cell-MRAM), and DW based MRAM and racetrack memories are elaborated in detail with respect to their architecture, performance, power consumption, and scalability.

Problems

Multiple Choice

1. **The disadvantage of SRAM over DRAM is**

   a. Integration density
   b. Reliable
   c. No periodic refresh
   d. Operational speed

2. **The volatile memory is**

    a. ReRAM
    b. STT-MRAM
    c. SRAM
    d. DW-MRAM

3. **Among the emerging memory technologies, the non-volatile memory with highest integration density is**

    a. ReRAM
    b. FeRAM
    c. STT-MRAM
    d. SOT-MRAM

4. **In Re-RAM, the two resistance states are represented by**

    a. Filament formation and destruction
    b. Parallel and anti-parallel configuration of the magnetic layers
    c. Crystalline and amorphous states of a material
    d. None of the above

5. **The material used for the phase change memory is**

    a. NiOx
    b. HfOx
    c. Chalcogenide
    d. TiOx

Answer Keys: 1-a, 2-c, 3-b, 4-a, 5-c

Short Answers

1. Explain the necessity of the novel memory technologies for the futuristic ultra-low computing applications.
2. Enlist the emerging non-volatile memory technologies.
3. Describe the single bit architecture of PCM.
4. Explain the advantages of spin-torque based MRAM over external field based MRAM.
5. Prepare a comparison table for emerging memory technologies.

References

1. A. Nitayama, Y. Kohyama, and K. Hieda, "Future directions for DRAM memory cell technology," *Int. Elect. Dev. Meet. 1998. Tech. Dig. (Cat. No.98CH36217)*, pp. 355–358, 1998.
2. M. Qazi, M. E. Sinangil, and A. P. Chandrakasan, "Challenges and directions for low-voltage SRAM," *IEEE Des. Test Compu.*, vol. 28, no. 1, pp. 32–43, 2011.

3. P. Gepner and M. F. Kowalik, "Multi-Core Processors: New way to achieve high system performance," *Int. Symp. Para. Compu. Elect. Eng.*, pp. 0–4, 2006.
4. S. Yoo, "Introduction to flash memory operation," *Proc. of IEEE*, vol. 91, no. 4, pp. 1–16, 2009.
5. C. J. Xue, Y. Zhang, Y. Chen, G. Sun, J. J. Yang, and H. Li, "Emerging non-volatile memories," *Proc. 17th IEEE/ACM/IFIP Int. Conf. Har./sof. Codes. Sys. Synth. - CODES + ISSS '11*, p. 325, 2011.
6. H.-S. P. Wong, S. Raoux, S. Kim, J. Liang, J. P. Reifenberg, B. Rajendran, M. Asheghi, and K. E. Goodson, "Phase change memory," *Proc. IEEE*, vol. 98, no. 12, pp. 2201–2227, 2010.
7. H. Akinaga and H. Shima, "Resistive random access memory (ReRAM) based on metal oxides," *Proc. IEEE*, vol. 98, no. 12, pp. 2237–2251, 2010.
8. S. Park, B. Magyari-kope, and Y. Nishi, "First-principles study of resistance switching in rutile TiO2 with oxygen vacancy," *Nonvol. Mem. Tech. Symp. 2008*, no. c, pp. 2–6, 2008.
9. Q. Liu, S. Long, W. Wang, Q. Zuo, S. Zhang, J. Chen, and M. Liu, "Improvement of resistive switching properties in ZrO_2-based ReRAM with implanted Ti ions," *IEEE Elect. Dev. Lett.*, vol. 30, no. 12, pp. 1335–1337, 2009.
10. H. Akinaga and H. Shima, "ReRAM technology; challenges and prospects," *IEICE Elect. Exp.*, vol. 9, no. 8, pp. 795–807, 2012.
11. U. Bottger and S. R. Summerfelt, "Ferroelectric random access memories," *Nanoelect. Inf. Tech.*, vol. 12, no. 10, pp. 565–590, 2003.
12. T. Mikolajick, S. Müller, T. Schenk, E. Yurchuk, S. Slesazeck, U. Schröder, S. Flachowsky, R. Van Bentum, S. Kolodinski, P. Polakowski, and J. Müller, "Doped Hafnium oxide – An enabler for ferroelectric field effect transistors," *Adv. in sci. and Tech.*, vol. 95, pp. 136–145, 2014.
13. J. M. Slaughter, "Recent advances in MRAM technology," *65th DRC Dev. Res. Conf.*, vol. 42, no. August 2006, pp. 245–246, 2006.
14. D. Apalkov, A. Ong, A. Driskill-Smith, M. Krounbi, A. Khvalkovskiy, S. Watts, V. Nikitin, X. Tang, D. Lottis, K. Moon, X. Luo, and E. Chen, "Spin-transfer torque magnetic random access memory (STT-MRAM)," *ACM J. Emer. Tech. Comp. Sys.*, vol. 9, no. 2, pp. 1–35, 2013.
15. X. Fong, Y. Kim, K. Yogendra, D. Fan, A. Sengupta, A. Raghunathan, and K. Roy, "Spin-transfer torque devices for logic and memory: Prospects and perspectives," *IEEE Trans. Compu. Des. Inte. Cir. Sys.*, vol. 35, no. 1, pp. 1–22, 2016.
16. T. Kawahara, K. Ito, R. Takemura, and H. Ohno, "Spin-transfer torque RAM technology: Review and prospect," *Microelect. Reliab.*, vol. 52, no. 4, pp. 613–627, 2012.
17. F.J.A.D. Broeder, W. Hoving, and P.J.H. Bloemen, "Magnetic anisotropy of multilayers," *J. of Magn. and Mag. Mat.*, vol. 93, pp. 562–570, 1991.
18. G. Prenat, K. Jabeur, P. Vanhauwaert, G. Di Pendina, F. Oboril, R. Bishnoi, M. Ebrahimi, N. Lamard, O. Boulle, K. Garello, J. Langer, B. Ocker, M. C. Cyrille, P. Gambardella, M. Tahoori, and G. Gaudin, "Ultra-fast and high-reliability SOT-MRAM: From cache replacement to normally-Off computing," *IEEE Trans. Mul. Compu. Sys.*, vol. 2, no. 1, pp. 49–60, 2016.
19. J. E. Hirsch, "Spin Hall Effect," *Phy. Rev. Let.*, vol. 83, no. 9, pp. 1834–1837, 1999.
20. H. Numata, T. Suzuki, N. Ohshima, S. Fukami, K. Nagahara, N. Ishiwata, and N. Kasai, "Scalable cell technology utilizing domain wall motion for high-speed MRAM," IEEE symp. of VLSI Tech., June 2007, vol. 89, pp. 232–233.

Chapter 2
Next Generation 3-D Spin Transfer Torque Magneto-resistive Random Access Memories

Spin transfer torque magneto-resistive random access memories (STT MRAMs) are non-volatile memories that potentially demonstrate high speed and integration density. These exclusive features of STT MRAMs are rapidly gaining attention of memory designers. They are strong contenders for futuristic embedded memory applications. However, further reduction in write power dissipation and cell size is essential to employ STT MRAMs for embedded applications. In most memory technologies, the select device has been the bottleneck towards increasing the array density [1]. Keeping this in mind, several novel architectures for select devices have been proposed by researchers working in the area [2–4]. They have been primarily focusing on vertical select devices to reduce the cell area. Kawahara et al. [5] suggested the possibility of a reduction in the cell area of spin transfer torque STT MRAMs down to $4F^2$ (F is the feature size) with a vertical select device. However, till date, the vertical select device could not be used in STT MRAM cells due to the requirement of high threshold switching current in the in-plane magnetic tunnel junction (MTJ) technology. The typical range of threshold switching current for conventional in-plane MTJs is 200–1200 µA [6, 7]. Such high current drive could not be achieved with minimum sized transistors, and hence, scaling towards $4F^2$ array density per cell was not feasible for STT MRAMs with in-plane MTJs. However, with the evolution of perpendicular magnetic anisotropy (PMA) MTJs which exhibit switching threshold of less than 100 µA [6], one can see decent prospects for higher array density in STT MRAMs.

In this chapter, the novel design of STT MRAMs on vertical silicon nano-wire (NW) platform is investigated. The chapter also discusses the architecture, operation and performance parameters of conventional STT MRAMs. Further, using extensive technology computer aided design (TCAD) and Hewlett simulation program with integrated circuit emphasis (HSPICE) simulations, the performance of STT MRAM with vertical gate-all-around (GAA) device is analyzed using a case study.

© The Author(s) 2017
B.K. Kaushik et al., *Next Generation Spin Torque Memories*,
SpringerBriefs in Applied Sciences and Technology,
DOI 10.1007/978-981-10-2720-8_2

The chapter comprises of six sections including the current introductory section. The architecture and functionality of the conventional STT MRAM cell are presented in Sect. 2.1. Further, Sect. 2.2 compares the cell size in different volatile and non-volatile memory technologies. The Sect. 2.3 outlines the next generation $4F^2$ STT MRAM and the simulation framework used for analyzing STT MRAM cells. Section 2.4 presents a case study on the design of the next generation STT MRAM cell with vertical GAA select device. The case study includes a comprehensive TCAD analysis of the proposed buried source GAA device followed by analysis and comparison of the bit cell simulation results on HSPICE using calibrated Verilog-A models. A fabrication methodology of proposed STT MRAM cell on vertical silicon nano-wire (NW) platform is presented in Sect. 2.5. Finally, Sect. 2.6, concludes the chapter.

2.1 Overview of Conventional STT MRAM: Architecture and Operation

STT MRAMs have 3D integration of MTJs with conventional CMOS technology. The memory cell of STT MRAMs has one MTJ and one NMOS select device, abbreviated as a 1T-1MTJ cell. The structure and properties of a typical 1T-1MTJ STT MRAM cell can be understood through Fig. 2.1. The bottom layer (pinned or fixed layer) of an MTJ has fixed magnetization due to its comparatively high magnetic coercivity. The top layer of the MTJ is known as free/recording layer whose magnetization can be switched by the spin torque acting on it. This spin torque is generated by the electric current, which is spin polarized by the pinned bottom layer. The directions of current for writing 1 and 0 are shown in the Fig. 2.1. A current from bit line (BL) to source line (SL) would make the magnetization orientation as P (to write a 0). Conversely, a current from SL to BL would make the magnetization states of two layers as AP (to write a 1). Besides this, for both the directions, the write currents need to be above a minimum threshold value for

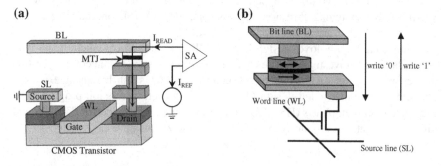

Fig. 2.1 **a** Typical integration of MTJ with CMOS technology in STT MRAM. **b** Write operation in STT MRAMs

proper switching of the MTJ. The minimum threshold value of current required for P(AP) to AP(P) switching is represented as $I_{LH0}(I_{HL0})$.

The data stored in MTJ is read by comparing current through MTJ with a reference resistance as shown in Fig. 2.1a. An optimum read voltage at BL is found by simultaneous consideration of *TMR* degradation effect with MTJ bias voltage and read current difference between P and AP states of MTJ [8]. Since, this read current has to be compared with a reference current to read the data, the difference in cell current between 0 and 1 stored cells should be high enough to be discernible.

In an MTJ, an interaction of spin polarized electrons with the local magnetic moment of FM layer takes place during which exchange of the spin angular momentum prompts to magnetization switching. For anti-parallel (AP or R_{AP}) to parallel (P or R_P) switching, the NMOS facilitates a current flow from the BL to SL as shown in Fig. 2.2a. Here, R_{AP} and R_P represent the parallel and anti-parallel resistance of an MTJ, respectively. The pinned FM layer acts as a spin filter, thus, producing a higher density of majority spin-polarized electrons. The spin-polarized electrons will sustain their spin polarization while crossing the tunnel barrier to finally exert STT on the free layer and decide its final state.

For P to AP switching, current flows from SL to BL (see Fig. 2.2a) during which electrons flowing through the tunnel barrier are polarized in the direction of free layer. On reaching the pinned layer, the electrons with spins in the direction of pinned layer pass through, while others are reflected back to the free layer. These reflected electrons exert STT on the free layer to decide its eventual magnetization state. Since these reflected electrons are fewer in number, MTJ shows an asymmetric behavior such that the threshold switching current density for P to AP switching is larger than AP to P change. Moreover, during P to AP switching, the NMOS operates in source follower mode such that the gate to source voltage (V_{GS}) is less than the supply voltage (V_{DD}). Low V_{GS} reduces the drive current which is also known as source degeneration [9].

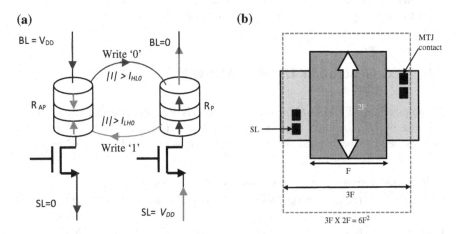

Fig. 2.2 **a** MTJ switching in STT MRAMs. **b** Layout of STT MRAM cell

Further, the Fig. 2.2b shows the layout of a typical 1T-1MTJ STT MRAM cell with a cell size of $6F^2$. The size of the cell depends on the minimum switching threshold current and current drivability of NMOS transistor. The major challenges associated with STT-MRAMs are: first, switching current reduction is crucial for achieving both high memory density and reducing the overall power consumption, second, sustaining high thermal stability for long period data retention [5].

2.2 Cell Size in Memories

The area required to store a single bit is known as cell size in memories. A comparison between different memory technologies is presented in Table 2.1 in terms of cell size [10]. The hard disk drives (HDDs) have the minimum cell size or highest integration density. Conventionally, the select device dominates the overall area occupied by a cell in most memory technologies [1]. International technology roadmap for semiconductors (ITRS) 2011 guidelines on memory design suggest a two terminal or vertical switch is the solution to growing needs of on-chip cache memory [1]. Researchers have proposed $6F^2$ cell with planar select devices and $4F^2$ cell with vertical select devices for various volatile and non-volatile memory technologies [2–4]. $4F^2$ cell on vertical silicon nano-wire (NW) platform has been demonstrated in various memory technologies. The $4F^2$ 1T-1R RRAM cell has been fabricated on vertical silicon NW platform by [4, 11–13]; wherein, the resistive random access memory (RRAM) stack is patterned to the same feature size above vertical GAA NW select device. 3D NAND flash memory of cell size $4F^2$ and beyond has been demonstrated by Kwong et al. [14]. Kawahara et al. [5] analyzed memory cell scalability for STT MRAMs for various transistor gate widths and informed the possibility of cell area reduction down to $4F^2$ with a vertical select device. However, no thorough analysis has been presented for vertical select device driving STT MRAM cells. Hence, experiments need to be initiated for realizing $4F^2$ architecture in the field of STT MRAMs too.

2.3 Next Generation $4F^2$ STT MRAM

The proposal for STT MRAM on a vertical silicon NW platform leads to the smallest cell area of $4F^2$. In logic applications with GAA devices, the source and gate contacts need to be created for each device. However, in the case of memory

Table 2.1 Comparison of cell size in different memory technologies

Technology	DRAM	SRAM	NAND flash	Hard disk drives	RRAM	STT MRAM
Cell size in (F^2)	6–12	140	1–4	2/3	4	20–60

applications, the source and gate terminals can be interconnected underneath. Here, the contacts need to be created after each data word. Thus, the proposed STT MRAM architecture with buried source line and word line saved significant area. Being circular in shape, the PMA MTJs can be stacked above the select device to save area. The planar MOSFETs have a saturation drive current per unit width of 900 µA/µm for high performance logic [15]. On the other hand, GAA MOSFETs have been reported to have a much higher saturation drive current of 2.6–2.9 mA/µm per unit diameter [16]. Hence, GAA transistor with the same diameter should provide a larger drive current. A vertical GAA transistor can act as an ideal select device that can provide sufficient drive current for efficient switching of an MTJ.

2.3.1 Proposed Architecture

The proposed architecture consists of a PMA MTJ stacked above the vertical GAA NMOS transistor is shown in Fig. 2.3. The diameter of the MTJ stack and silicon NW is F. The overall area of each cell is $4F^2$ as shown in Fig. 2.3. It has a buried n or $n+$ layer above the oxide (SiO_2) or p-Si substrate, which can be formed after the formation of p-type NW. This forms the buried source line (SL). The SL and word line (WL) contacts are brought out through vias. The structure of the vertical GAA NMOS is shown in Fig. 2.3b. For SL = V_{DD} and bit line (BL) = 0 condition, the vertical GAA transistor will be operating with the bottom-as-drain and top-as-source. Conversely, for BL = V_{DD} and SL = 0, the GAA transistor operates with top-as-drain and bottom-as-source (see Fig. 2.3a).

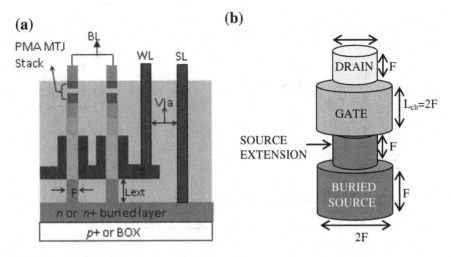

Fig. 2.3 **a** The architecture of vertical silicon NW 3D STT MRAMs. **b** The structure of vertical GAA NMOS select device

2.3.2 Performance Parameters and Windows

The performance parameters of STT MRAM can be well understood from Fig. 2.4 that demonstrates hysteretic behavior, cell current, and read/write windows. STT MARM write (read) operation is characterized by a write (read) margin and write (read) error rate. The write margins are defined as the difference between the write current and the switching threshold current [17]. Write error rate (WER) is the probability of failure during a particular write operation.

Read disturb rate (RDR) is the probability of accidental write during a read operation. Both, WER and RDR are quantified using the ubiquitous formalism of stochastic STT switching of a nano-magnetic layer at a finite temperature [18]. The probability of switching at a particular current ($I_{write/read}$), whether read or write, is expressed as,

$$P_{SW} = 1 - \exp\left\{ -\frac{t_{PW}}{\tau_0}\exp\left[-\frac{E}{k_BT}\left(1 - \frac{I_{write/read}}{I_{C0}}\right)\right]\right\} \tag{2.1}$$

where, t_{PW} is the pulse width of read/write current, τ_0 is the inverse of the attempt frequency, I_{C0} is the switching threshold current (either I_{LH0} or I_{HL0}), E/k_BT is the thermal stability factor for an MTJ. Using this expression, the WER and RDR are expressed as

$$WER = 1 - P_{SW}, RDR = P_{SW} \tag{2.2}$$

WER decreases with an increase in write current above the threshold current. For reliable write (read) operation in STT MRAM, WER (RDR) should be less than

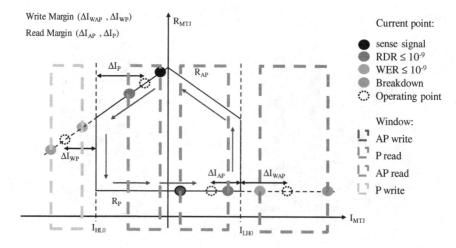

Fig. 2.4 Demonstrating resistance current hysteresis loop with read and write window

10^{-9} and thus sets the minimum (maximum) allowed value of write (read) current. The upper limit on write current is set by thin dielectric layer's breakdown voltage.

In conventional STT MRAM cells, the supply voltage (V_{DD}) is constant at a particular technology node. The minimum cell size in STT MRAM is determined by the smaller of the two write margins WM_P and WM_{AP}. The width of the access device is increased to achieve an appropriate value of overall WM. However, in order to maintain the overall area of GAA based STT MRAM cell to $4F^2$, the diameter of nano-wire has been kept constant. In addition, the vertical nano-wire technology does not allow the same flexibility of changing the dimensions as the conventional CMOS technology. Therefore, it is more suitable that the supply voltage V_{DD} is varied to achieve the optimum write current in vertical NW based STT MRAM.

2.3.3   Simulation Framework

Till date, STT MRAMs have been realized using planar CMOS technology only. The minimum attainable single cell size with a planar n-channel metal oxide semiconductor (NMOS) as access/select device is 6–8F^2. Using vertical NMOS as a select device, the cell area can be reduced to $4F^2$ [1, 5]. Since, calibrated SPICE models are not available for such novel structured devices; therefore, their impact on MRAMs cannot be analyzed on SPICE. It is advisable to use mixed mode device-circuit simulation framework shown in Fig. 2.5 to understand the effect of device design on MRAMs and non-volatile hybrid MTJ based logic circuits. TCAD device simulator is used to obtain the performance of an arbitrary select device such

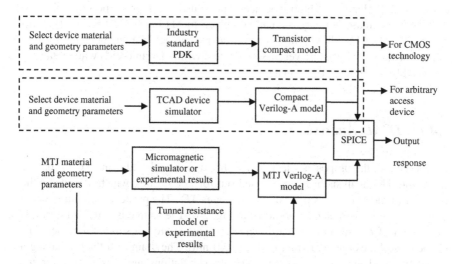

Fig. 2.5 SPICE based mixed mode simulation framework with Verilog-A MTJ model

as a vertical GAA transistor. Here, a micromagnetic simulator can also be used for calculating threshold switching current for an MTJ [19]. The simulation platform used is decided by a tradeoff between computation time and accuracy.

2.4 Case Study

In this section, using a case study the performance of STT MRAM with vertical select devices is analyzed. A comparison with conventional planar STT MRAM cell is also presented. The proposed STT MRAM cell is analyzed using calibrated Verilog-A models for PMA MTJ and vertical GAA NMOS transistor (BSIM CG) by HSPICE simulations. Write/dynamic power dissipation will be same for both vertical GAA and planar STT MRAM cells while using SiO_2 gate dielectric (GD) based select device. Hence, in the second part of the case study a superior 3-dimensional (3D) vertical silicon nano-wire (NW)/GAA with high-k (HfO_2) gate dielectric device based STT MRAM is presented. A significant improvement in write power can be achieved using high-k gate GAA device, as it provides higher drive current capability at lower supply voltage (V_{DD}). In addition, the implementation of high-k gate dielectric does not have any explicit impact on the MTJ switching time; and hence, does not affect the delay performance in STT MRAMs and MTJ based hybrid spintronic-CMOS circuits. In contrast to this, the conventional CMOS circuits follow the standard inverse relationship between delay and drive current, wherein, the propagation delay decreases continuously with an increase in drive current [20]. Using high-k GDs in CMOS circuits, the increased gate capacitive delay annuls the effect of increase in the drive current. In STT MRAMs, an inherent MTJ magnetization switching delay of 0.1–1 ns is inevitably present for all practical write current densities [21]. Hence, the high-k devices can be used to improve power savings as long as the increase in delay is small compared to MTJ switching delay. The proposed STT MRAM cell with high-k select device can achieve an appreciably larger tradeoff window between power dissipation and write margin (WM) while retaining appreciably good delay performance.

2.4.1 TCAD Analysis

In this section, the proposed select device structure is analyzed using TCAD device simulator. The analysis has been carried out for $F = 40$ nm, which is same as the diameter of the MTJ to keep the overall cell area $4F^2$. The proposed structure with a buried source of diameter $2F$ is shown in Fig. 2.6. The source is extended upwards to a height of $F = 40$ nm so that the gate and source are not shorted together. The diameter of the extension region is also F (40 nm). The drain is at the top having a diameter and length of F (40 nm). Device simulations are carried out for the

(a) **(b)**

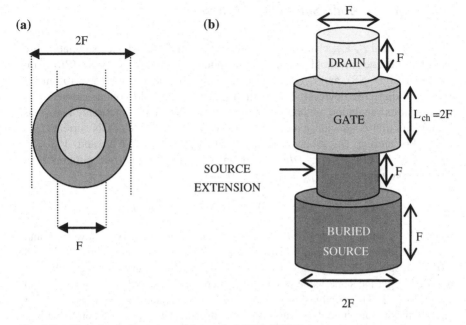

Fig. 2.6 **a** Top view **b** front view of the vertical GAA NMOS device

proposed GAA structure on TCAD [22] for gate/channel lengths (L_{ch}) of 40 (F), 80 ($2F$), and 120 nm ($3F$). The buried substrate layer is not shown for simplicity.

Source, drain, and source extension regions are heavily doped with a uniform n-type doping concentration of 1×10^{20} cm^{-3}. The channel is uniformly doped with a p-type doping concentration of 1×10^{16} cm^{-3}. The work function and gate oxide thickness of the cylindrical gate are 4.61 eV and 2 nm, respectively. The heavily doped extension region ensures a high current drive. The gate dielectric and feature size used during simulations are mentioned, specifically, when changed for the select devices.

2.4.2 TCAD Simulation Setup

TCAD device simulator self-consistently solves the Poisson and carrier continuity equations. Fermi-Dirac and band-gap narrowing based carrier statistics model are activated to account for carrier transport with heavily doped source/drain regions. Besides this, the Lombardi CVT (concentration, voltage, temperature) mobility model is used to accounts for all common effects related to carrier mobility in non-planar devices. The dependence of carrier mobility on impurity concentration, transverse electric field, longitudinal electric field, velocity saturation, and temperature are captured using the Lombardi's model. Furthermore, concentration-dependent Shockley-Read-Hall (SRH) recombination/generation model is also included in the TCAD simulation setup.

2.4.3   Mixed-Mode Simulation Results

The I_D-V_{DS} and I_D-V_{GS} characteristics obtained from TCAD device simulation are used to calculate threshold voltage, drain induced barrier lowering (DIBL), sub-threshold slope (SS), and I_{ON}/I_{OFF} for L_{ch} of 40, 80, and 120 nm (F = 40 nm for SiO$_2$ gate dielectric devices). The method used for threshold voltage extraction is "the linear extrapolation method in the linear region" which is commonly known as "maximum transconductance method" [23, 24]. The magnitude of ON current (I_{ON}) and OFF current (I_{OFF}) are calculated at $V_{GS} = V_{DS}$ = 1.6 V and V_{GS} = 0, V_{DS} = 1.6 V, respectively. DIBL is defined as the normalized difference in threshold voltages when V_{DS} is changed from V_{DS_lin} to V_{DS_sat}.

$$DIBL = (V_{t\_lin} - V_{t\_sat})/(V_{DS\_sat} - V_{DS\_lin}) \qquad (2.3)$$

where, V_{t_lin} and V_{t_sat} are the threshold voltage in the linear (very low V_{DS}) and saturation region, respectively. The values of V_{DS_sat} and V_{DS_lin} are 1.6 V and 0.05 V, respectively. V_{t_sat} is the value of V_{GS} on I_D-V_{GS} curve for $V_{DS} = V_{DS\_sat}$ required to get the same value of current, which is obtained when $V_{GS} = V_{t\_lin}$ and $V_{DS} = V_{DS\_lin}$. The subthreshold slope (SS) is the change in V_{GS} required for altering the subthreshold drain current by one decade (10 times). The corresponding results are shown in Table 2.2. Although, the device with 40 nm gate length has the largest current drive, but it severely suffers from short channel effects. The DIBL, OFF current, and subthreshold slope parameters are comparatively large for the device with 40 nm gate length. It is because of lower electrostatic gate control at smaller gate length. Evidently, the device with 120 nm gate length demonstrates the best performance. The smaller gate length would also mean higher Joule heating in a smaller volume. Hence, larger gate length should be preferred for the vertical GAA architecture in STT MRAM.

In order to compare the performance of GAA based cell with the conventional STT MRAM cell, TCAD simulations are carried out for planar NMOS transistor also. The gate work function and oxide thickness of the planar NMOS are 4.61 eV and 1.5 nm, respectively. However, the source, drain, and channel doping are same as that for GAA NMOS. The performance parameters calculated for a planar NMOS with different device dimensions are placed in Table 2.3. The performance of the planar NMOS is poor, especially for small gate-length, due to large short channel effects that are measured at a high V_{DD} of 1.6 V. Although, the short channel effects of the planar NMOS can be reduced by decreasing the source and

Table 2.2 TCAD results of SiO$_2$ GD GAA device for different gate lengths

L_{ch} (nm)	V_{t_lin} (V)	DIBL (mV/V)	I_{ON} (μA)	I_{ON}/I_{OFF}	SS (mV/decade)
40	0.279	115.0	214.6	3.8×10^3	103.0
80	0.300	22.1	194.0	3.3×10^6	63.4
120	0.307	13.6	175.5	1.5×10^7	60.3

Table 2.3 TCAD results of planar NMOS device with SiO_2 GD

W_{NMOS} (nm)	L_{ch} (nm)	V_{t_lin} (V)	DIBL (mV/V)	I_{ON} (μA)	I_{ON}/I_{OFF}	SS (mV/decade)
40	80	0.250	354.0	85.8	18.3	768.3
40	120	0.275	280.0	72.8	142.8	294.2
80	120	0.275	310.0	145,6	142.7	294.2

Fig. 2.7 Comparison of the I_D-V_{GS} characteristics of GAA (40 nm nano-wire diameter and L_{ch} = 120 nm) and planar (W_{NMOS} = 80 nm and L_{ch} = 120 nm) NMOS with SiO_2 GD

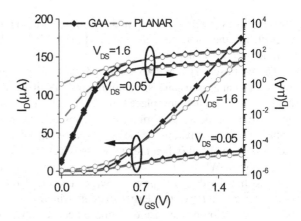

drain doping concentration, but that will reduce I_{ON} also. A comparison of I_D-V_{GS} characteristic in Fig. 2.7 confirms that GAA NMOS has much lower OFF current than the planar NMOS devices.

For SL = V_{DD} and BL = 0, the vertical GAA transistor would be operating with the bottom-as-drain and top-as-source. Therefore, the I_D-V_{DS} characteristics of the proposed structure should also be analyzed with the bottom-as-drain (buried source in Fig. 2.6). However, Fig. 2.8 clearly shows a minute difference between $|I_D|$-$|V_{DS}|$

Fig. 2.8 Comparison of $|I_D|$-$|V_{DS}|$ characteristics with top-as-drain and bottom-as-drain operation for the GAA NMOS of 40 nm diameter (SiO_2 GD) with L_{ch} = 120 nm

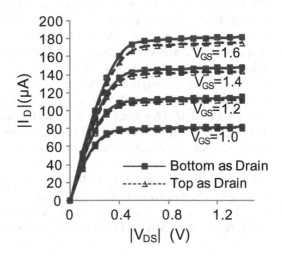

Table 2.4 Comparison between proposed cell and $10F^2$ cell with planar NMOS (at V_{DD} = 1.6 V, F = 40 nm and SiO_2 GD)

Select device	WM_P (μA)	WM_{AP} (μA)	t_p (ns)	t_{AP} (ns)	Power (pW) WL = 0 V	Power (μW) WL = 1.6 V
GAA	23.3	34.0	0.43	0.25	16.2	112.4
Planar	17.6	29.0	0.52	0.32	1140.0	107.0

characteristics under the two modes of operation. This difference can be safely neglected in the subsequent analysis and the device can be considered to be having symmetric *I-V* characteristics.

The proposed STT MRAM cell and conventional cell with planar NMOS are analyzed using transient simulations on HSPICE using calibrated Verilog-A models. The results are placed in Table 2.4. Undoubtedly, the proposed cell demonstrates a better performance in terms of power dissipation and WMs. The leakage power dissipation for the proposed cell is 3 orders of magnitude lower than the conventional cell (when the cell is not selected for writing). The dynamic power dissipation at WL = 1.6 V (V_{DD}) is higher, although, here, the point of consideration is that the dynamic power always has a tradeoff with the P and AP WMs (write currents). There is a larger tradeoff window between the WM and dynamic power dissipation in the case of proposed cell with GAA device. The operation of proposed GAA based cell is also shown in Fig. 2.9 using a timing diagram.

2.4.4 Impact of High-k GAA Devices

Till now, the STT MRAM cells with vertical GAA device use SiO_2 GD. However, to fully explore the impact of all around gate control of vertical GAA devices on STT MRAM cells, HfO_2 based devices with high dielectric constant of 22 need to be used. In this section, the impact of high-k (HfO_2) GAA select device on the performance of an STT MRAM cell is analyzed. In addition, its performance is compared to STT MRAM cell using GAA devices with the conventional SiO_2 gate dielectric. Initially, TCAD simulations are carried out from feature size of 40–70 nm, while taking into account the mobility degradation effect in high-k (HfO_2) GD devices [25]. The mobility degradation is benchmarked using data obtained from Chau et al. [26] for metal gate devices. The work function and gate dielectric thickness are considered as 4.61 eV and 2 nm, respectively. Compact Verilog-A models are developed and calibrated individually for the HfO_2 GD based vertical GAA transistor and PMA MTJ.

Berkeley short-channel IGFET (Insulated gate field effect transistor) model of combined multi-gate (BSIM CMG) [27] is calibrated based on the results acquired from TCAD device simulations. Moreover, experimental results [28, 29] for high-performance PMA MTJs from Table 2.5 are used to calibrate compact

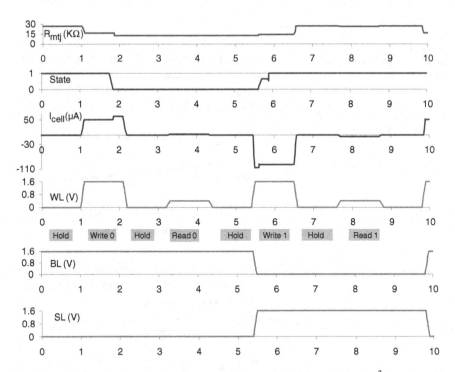

Fig. 2.9 A timing diagram showing the successful operation of the proposed $4F^2$ STT MRAM cell at $V_{DD} = 1.6$ V

Table 2.5 Experimental results for PMA MTJs

MTJ stack	Diameter (F in nm)	TMR (%)	RA (Ω-μm^2)	STT efficiency ($k_BT/\mu A$)
stack-1 [28]	40–70	102	6.6	1.3–3.0
stack-2 [29]	30–70	150	12	2.2–3.8

*STT efficiency is the ratio of $\Delta = E/k_BT$ in units of k_BT and I_{C0}, where I_{C0} is the average of (I_{LH0} and I_{HL0}) and Δ is the Boltzmann's stability factor. RA and TMR are the resistance area product and tunnel magneto-resistance of an MTJ, respectively.

Verilog-A models corresponding to low RA stack-1 and high RA stack-2. Here, STT efficiency is the ratio of Δ (= E/k_BT in units of k_BT) and I_{C0}, where, I_{C0} is the average of (I_{LH0} and I_{HL0}), and Δ is the Boltzmann's/thermal stability factor.

Bit cell level HSPICE simulations are carried out using calibrated compact models at different values of V_{DD} and F. The pulse width of write current lies between 2 and 50 ns so as to provide sufficient time for switching at given F. The write currents and threshold switching currents (I_{LH0} and I_{HL0}) for different feature sizes ranging from 40 to 70 nm are shown in Fig. 2.10a, b when SYM HfO$_2$ and SiO$_2$ based GAA devices are used in STT MRAM cells. It is clearly observed that

Fig. 2.10 Comparison between the write current obtained for STT MRAM cells with high-*k* (HfO$_2$) and SiO$_2$ GD vertical GAA NMOS devices **a** low *RA* MTJ stack-1 used during simulation **b** high *RA* stack-2 used during simulation

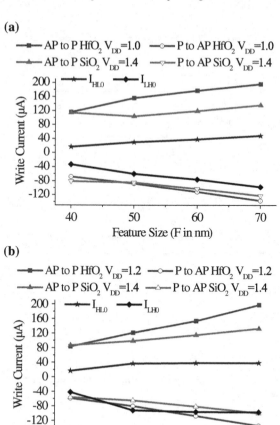

for all *F* between 40 and 70 nm, the high-*k* devices can be employed to achieve high WMs at lower V_{DD} (by 0.4 V). Although the write current for P to AP switching are overlapping for SiO$_2$ and high-*k* cells, but the analysis of SiO$_2$ cells was carried out at a higher V_{DD} of 1.4 V. Hence, high-*k* devices can reduce power dissipation in STT MRAMs by V_{DD} reduction.

2.4.5 Impact of High-k GD on Delay

The additional delay introduced due to the use of select devices with high-*k* gate dielectric is compared with the switching time of MTJ. The introduction of the high-*k* dielectric increases the WL capacitance. Elmore delay model [20] is used to determine the delay from the write line driver node to the gate electrode of a

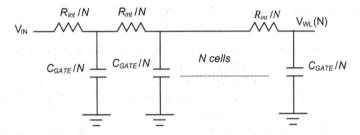

Fig. 2.11 Equivalent RC model for the 3D NW STT MRAM architecture

particular select device. If C_{GATE}/N is the gate capacitance and R_{int}/N is the interconnect resistance between cells (see Fig. 2.11), the propagation delay (τ_D) at WL after N cells is expressed as,

$$\tau_D = R_{int} C_{GATE} \left(\frac{N+1}{N} \right) \tag{2.4}$$

The propagation delay of N (N = 100 and 200) cells for different gate dielectrics is compared with the typical switching time of PMA MTJ at current of 100 μA. The percentage WL delay normalized by the MTJ switching time is plotted in Fig. 2.12. For N = 100 cells the delay of HfO_2, SiO_2 and Si_3N_4 is less than 10% of the switching time required by a PMA MTJ. Hence, high-k gate dielectric select devices need to be analyzed thoroughly, because they could lead to V_{DD} reduction at the cost of increase in WL capacitance. As, this delay due to the increase in WL capacitance can be considered nominal with respect to MTJ switching time.

Fig. 2.12 Percentage switching delay of WL after N cells for various dielectrics with respect to MTJ switching delay

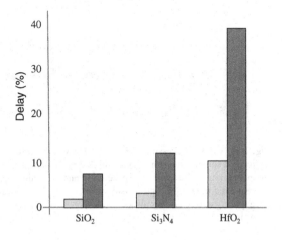

2.5   Proposed Fabrication Methodology

Several research groups have successfully fabricated $4F^2$ cell with vertical select devices for various memory technologies [11, 12, 14] on vertical silicon nano-wire platform. The 1T-1-resistor resistive random access memory (RRAM) cell has been fabricated on vertical silicon nano-wire (NW) platform by [4, 11–13] wherein the RRAM stack is patterned to the same feature size above vertical GAA NW select device. $4F^2$ 3D NAND flash and junction-less silicon-oxide-nitride-oxide-silicon (JL-SONOS) memory have been demonstrated in Kwong et al. [14]. Based on the methodology given in the aforementioned references, processing steps for fabricating $4F^2$ STT MRAM cell on vertical silicon NW platform are presented in Figs. 2.13 and 2.14.

The fabrication of MTJs should follow the same back-end-of-line (BEOL) semiconductor processing (see Fig. 2.13); wherein, the MTJ stack is deposited between successive metal interconnect layers after the front-end-of-line (FEOL) processing of the vertical GAA select device.

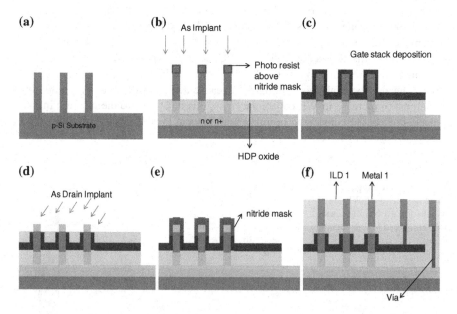

Fig. 2.13 FEOL Processing [13, 30, 31] **a** vertical pillar etch. **b** As implantation to form the source and source extension regions, high density plasma oxide deposition **c** gate stack deposition **d** top amorphous Si etched to expose drain and As drain implantation **e** nitride mask formation to prevent nano-pillar tip **f** high density plasma oxide deposition, chemical mechanical planarization for exposing and removing nitride mask, pre-metal dielectric oxide deposition, metallization with Al pads

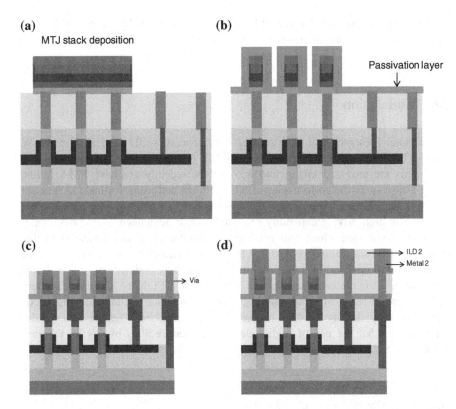

(a) MTJ stack deposition

(b) Passivation layer

(c) Via

(d) ILD 2 / Metal 2

Fig. 2.14 BEOL Processing [32] **a** formation of MTJ stack over planarized surface. The stack includes first electrode (generally Ta), a fixed layer, tunnel barrier, free layer and second electrode. The fixed and free layers are generally composed of 2–3 layers. **b** A single mask etching process and deposition of dielectric passivation layer for protection. **c** ILD-1 deposition that encapsulates MTJ stack followed by removal using planarization, and via formation. **d** Another dielectric passivation, ILD2 deposition, planarization to expose, and metallization to form metal-2 interconnect

The fabrication of MRAMs involves multiple steps of deposition and etching during the fabrication of MTJ stack. Especially during the etching of primary layers (ferromagnetic layers and tunnel barrier) constituting an MTJ, the sidewall redeposition of unwanted material on MTJs could cause failure of the device [33, 34] The etching process should be highly selective which is difficult to achieve at low dimensions and further difficulties arise because of the vertical GAA NW transistor. The two major challenges during fabrication of STT MRAM cell are:

- The etching process that goes from the top electrode to bottom electrode should have a high selectivity so that the MTJ cell is protected from any erosion during etching. Retaining the parameters like *TMR* and stability factor is of paramount importance during etching and pattering of MTJ stack.

- The redeposition material should be completely removed by the end of etching cycle as it can cause shorts across the tunnel barrier of MTJ.

2.6   Conclusion

This chapter presented and analyzed the next generation 3D vertical silicon NW based STT MRAMs with vertical GAA select device. The design tradeoffs and constraints are analyzed taking into account the recently available PMA MTJs. Using a case study the performance of STT MRAM cell using vertical GAA select device is analyzed and compared with conventional STT MRAM with planar select device. The high driving capability provided a clear advantage over conventional planar access device which was reflected in simulation results presented in this chapter. TCAD simulations proved that the GAA device provided a drive current of 2.8 mA/µm as opposed to 1.16 mA/µm by planar access device at $V_{DD} = 1.2$ V (SiO_2 GD). Thus the proposed vertical silicon NW STT MRAMs offer excellent performance in terms of write margins, power dissipation while achieving maximum 2D array density of $4F^2$. Moreover, the analysis for feature size 70 nm demonstrates that the use of high-k device can reduce V_{DD} by 0.4 V, thereby, reducing write power up to 25%.

Problems

Multiple Choice

1. **The main bottleneck for the scalability of STT-MRAM array is**

   a. Pinned layer
   b. Free layer
   c. Tunnel barrier
   d. Select device

2. **One of the ferromagnetic layers of the MTJ is pinned using the material with**

   a. Low coercivity
   b. High coercivity
   c. Low retaintivity
   d. High retaintivity

3. **The write margin (*WM*) increases with the**

   a. Increase in switching threshold
   b. Decrease in operating current
   c. Increase in operating current
   d. None of these

4. **The read disturb rate (*RDR*) decreases with**

    a. Decrease in switching threshold
    b. Increase in read current
    c. Increase in thermal stability factor
    d. All of these

5. **The overall delay of high-*k* GD based GAA device**

    a. Remains constant
    b. Decreases
    c. Increases
    d. None of these

6. **The purpose of source extension region is**

    a. To keep separate gate and buried source
    b. To make shorted gate and buried source
    c. To increase the source length
    d. To enhance the performance

7. **The overall delay of STT-MRAM using high-*k* GD based GAA device**

    a. Increases
    b. Remains constant
    c. Decreases
    d. None of these

8. **For an MTJ, Read disturb rate (RDR) is**

    a. The probability of false read operation during a write operation
    b. The probability of no write during read operation
    c. The probability of accidental write during a read operation
    d. None of the above

9. **For an MTJ, the width of the access device is increased to**

    a. Achieve an appropriate value of overall write margin
    b. To reduce the overall write margin
    c. To reduce the write error rate
    d. To reduce the heating effect

10. **The high-*k* gate GAA improves the write power due to**

    a. Low drive current at high voltage
    b. High drive current capability at low voltage
    c. High drive current at high voltage
    d. None of the above

Answer Keys: 1-d, 2-b, 3-c, 4-d, 5-c, 6-a, 7-b, 8-c, 9-a, 10-b

Short Answers

1. Explain the advantages of STT-MRAM for making it strong contender for futuristic embedded memory technology.
2. Describe the tunnel barrier reliability issues.
3. Enlist the advantages of high-k devices.
4. Analyze the experimental results for PMA MTJs.
5. Enlist the steps for FEOL processing.

References

1. Emerging research devices (2011). *International technology road map for semiconductors* [Online]. Available: http://www.itrs.net
2. T. Schloesser, F. Jakubowski, J. V. Kluge, A. Graham, S. Slesazeck, M. Popp, P. Baars, K. Muemmler, P. Moll, K. Wilson, A. Buerke, D. Koehler, J. Radecker, E. Erben, U. Zimmermann, T. Vorrath, B. Fischer, G. Aichmayr, R. Agaiby, W. Pamler, T. Schuster, W. Bergner, and W. Mueller, "6F^2 buried wordline DRAM cell for 40 nm and beyond," *Proc. IEEE Int. Elect. Dev. Meet. (IEDM 2008)*, San Francisco, CA, 2008, pp. 1–4.
3. H. Chung, H. Kim, H. Kim, K. Kim, S. Kim, K.-W. Song, J. Kim, Y. C. Oh, Y. Hwang, H. Hong, G.-Y. Jin, and C. Chung, "Novel 4F^2 DRAM cell with vertical pillar transistor (VPT)," *Proc. IEEE Eur. Sol. Dev. Res. Conf. (ESSDERC 2011)*, Helsinki, 2011, pp. 211–214.
4. Z. Fang, X. P. Wang, X. Li, Z. X. Chen, A. Kamath, G. Q. Lo, and D. L. Kwong, "Fully CMOS-compatible 1T1R integration of vertical nanopillar GAA transistor and oxide-based RRAM cell for high-density," *IEEE Trans. Elect. Dev.*, vol. 60, no. 3, pp. 1108–1113, 2013.
5. T. Kawahara, K. Ito, R. Takemura, and H. Ohno, "Spin-transfer torque RAM technology: review and prospect," *Microelect. Reliab.*, vol. 52, no. 4, pp. 613–627, 2012.
6. S. Ikeda, H. Sato, M. Yamanouchi, H. Gan, K. Miura, K. Mizunuma, S. Kanai, S. Fukami, F. Matsukura, N. Kasai, and H. Ohno, "Recent progress of perpendicular anisotropy magnetic tunnel junctions for nonvolatile VLSI," *Spin World Sci.*, vol. 02, no. 03, pp. 1240003-1–1240003-12, 2012.
7. S. Ikeda, J. Hayakawa, Y. M. Lee, F. Matsukura, and Y. Ohno, T. Hanyu, H. Ohno "Magnetic tunnel junctions for spintronic memories and beyond," *IEEE Trans. Elect. Dev.*, vol. 54, no. 5, pp. 991–1002, 2007.
8. T. Kawahara, R. Takemura, K. Miura, J. Hayakawa, S. Ikeda, Y. M. Lee, R. Sasaki, Y. Goto, K. Ito, T. Meguro, F. Matsukura, H. Takahashi, H. Matsuoka, and H. Ohno, "2 Mb SPRAM (spin-transfer torque RAM) with bit-by-bit bi-directional current write," *IEEE Trans. Sol. Cir.*, vol. 43, no. 1, pp. 109–120, 2008.
9. S. Verma, P. K. Pal, S. Mahawar, and B. K. Kaushik, "Performance Enhancement of STT MRAM using asymmetric-k sidewall-spacer NMOS," *IEEE Trans. on Elect. Dev.*, vol. 63, no. 7, pp. 2771–2776, 2016.
10. J. J. Yang, D. B. Strukov, and D. R. Stewart, "Memristive devices for computing.," *Nat. Nanotech.*, vol. 8, no. 1, pp. 13–24, 2013.
11. X. P. Wang, Z. X. Chen, X. Li, A. R. Kamath, L. J. Tang, D. Mei, Y. Lai, P. C. Lim, D. Teng, H. Li, N. Singh, P. Guo, Q. Lo, and D. Kwong, "HfOx-based RRAM cells with fully CMOS compatible technology," *Proc. IEEE Int. Conf. Sol. Inte. Cir. (ICSIC 2012)*, Singapore, 2012, pp. 1–6.

12. B. Chen, X. Wang, B. Gao, Z. Fang, J. Kang, L. Liu, X. Liu, G.-Q. Lo, and D.-L. Kwong, "Highly compact ($4F^2$) and well behaved nano-pillar transistor controlled resistive switching cell for neuromorphic system application," *Nat. Sci. Rep.*, vol. 4, pp. 6863-1–6863-5, 2014.

13. X. Wang, Z. Fang, X. Li, and B. Chen, "Highly compact 1T-1R architecture ($4F^2$ footprint) involving fully CMOS compatible vertical GAA nano-pillar transistors and oxide-based RRAM cells exhibiting excellent NVM properties and ultra-low power operation," *Proc. IEEE Int. Elect. Dev. Meet. (IEDM 2012)*, San Francisco, CA, 2012, pp. 20.6.1–20.6.4.

14. D.-L. Kwong, X. Li, Y. Sun, G. Ramanathan, Z. X. Chen, S. M. Wong, Y. Li, N. S. Shen, K. Buddharaju, Y. H. Yu, S. J. Lee, N. Singh, and G. Q. Lo, "Vertical silicon nanowire platform for low power electronics and clean energy applications," *J. Nanotechn.*, vol. 2012, pp. 1–21, 2012.

15. Process integration device and structures (2001). *International technology roadmap for semiconductors* [Online]. Available: http://www.itrs.net

16. Y. Song, Q. Xu, J. Luo, and H. Zhou, "Performance breakthrough in gate-all-around nanowire n- and p-type MOSFETs fabricated on bulk silicon substrate," *IEEE Elect. Dev. Lett.*, vol. 59, no. 7, pp. 1885–1890, 2012.

17. S. Verma, S. Kaundal, and B. K. Kaushik, "Novel $4F^2$ buried-source-line STT MRAM cell with vertical GAA transistor as select device," *IEEE Trans. on Nanotechn.*, vol. 13, no. 6, pp. 1163–1171, 2014.

18. Z. Li and S. Zhang, "Thermally assisted magnetization reversal in the presence of a spin-transfer torque," *Phys. Rev. B*, vol. 69, no. 13, p. 134416, Apr. 2004.

19. S. Verma, M. S. Murthy, and B. K. Kaushik, "All spin logic (ASL): A micromagnetic perspective," *IEEE Trans. on Mag.*, vol. 51, no. 10, pp. 3400710-1–3400710-7, 2015.

20. J. M. Rabaey, A. Chandrakasan, and B. Nikolic, Digital integrated circuits: A design perspective, Upper Saddle River, NJ, Prentice-Hall, 2003, pp. 177–233.

21. D. D. Tang, and Y. J. Lee, Magnetic memory fundamentals and technology, Cambridge, UK, Cambridge University Press, 2010, pp. 122–164.

22. ATLAS user's manual, Silvaco Inc., 2012. Available: www.silvaco.com

23. L. Dobrescu, M. Petrov, D. Dobrescu, and C. Ravariu, "Threshold voltage extraction methods for MOS transistors," *Proc. 23rd IEEE Int. Semicond. Conf. (CAS 2000)*, Sinaia, 2000, pp. 371–374.

24. A. Bazigos, M. Bucher, J. Assenmacher, S. Decker, W. Grabinski, and Y. Papananos, "An adjusted constant-current method to determine saturated and linear mode threshold voltage of MOSFETs," *IEEE Trans. Elect. Dev.*, vol. 58, no. 11, pp. 3751–3758, 2011.

25. W. J. Zhu, and T. P. Ma, "Temperature dependence of channel mobility in HfO_2-gated NMOSFETs," *IEEE Elect. Dev. Lett.*, vol. 25, no. 2, pp. 89–91, 2004.

26. R. Chau, J. Brask, S. Datta, G. Dewey, M. Doczy, B. Doyle, J. Kavalieros, B. Jin, M. Metz, A. Majumdar, and M. Radosavljevic, "Application of high-k gate dielectrics and metal gate electrodes to enable silicon and non-silicon logic nanotechnology," *Microelect. Eng.*, vol. 80, pp. 1–6, 2005.

27. V. Sriramkumar, N. Paydavosi, J. Duarte, D. Lu, C. Hsun Lin, M. Dunga, S. Yao, T. Morshed, A. Niknejad and C. Hu, "BSIM-CMG 107.0.0 multi-gate MOSFET compact model: technical manual," Dept. of Electrical Engineering and Computer Sciences, Univ. of California, Berkeley, 2013.

28. G. Jan, Y.-J. Wang, T. Moriyama, Y.-J. Lee, M. Lin, T. Zhong, R.-Y. Tong, T. Torng and P.-K Wang "High spin torque efficiency of magnetic tunnel junctions with MgO/CoFeB/MgO free layer," *Appl. Phys. Exp.*, vol. 5, pp. 093008-1–093008-3, 2012.

29. L. Thomas, G. Jan, J. Zhu, H. Liu, Y.-J. Lee, S. Le, R.-Y. Tong, K. Pi, Y.-J. Wang, D. Shen, R. He, J. Haq, J. Teng, V. Lam, K. Huang, T. Zhong, T. Torng, and P.-K. Wang, "Perpendicular spin transfer torque magnetic random access memories with high spin torque efficiency and thermal stability for embedded applications (invited)," *J. Appl. Phys.*, vol. 115, no. 17, pp. 172615-1–172615-6, 2014.

30. B. Yang, K. D. Buddharaju, S. H. G. Teo, N. Singh, G. Q. Lo, and D. L. Kwong, "Vertical silicon-nanowire formation and gate-all-around MOSFET," *IEEE Elect. Dev. Lett.*, vol. 29, no. 7, pp. 791–794, 2008.
31. R. Gandhi, Z. Chen, N. Singh, K. Banerjee, and S. Lee, "Vertical Si-nanowire *n*-type tunneling FETs with low subthreshold swing at room temperature,' *IEEE Elect. Dev. Lett.*, vol. 32, no. 4, 437–439, 2011.
32. S. H Kang, D. Bang, and K. Lee, "One-mask MTJ integration for STT MRAM," U.S. Patent 2009/0261433 A1, 2009.
33. E. J. O'Sullivan, Magnetic tunnel junction-based MRAM and related processing issues "*IBM Research Report*," RC23525, 2005.
34. Yuchen Zho, and Yiming Huai, "STT-MRAM manufacturing method with in-situ annealing," U.S. Patent US8758850 B2, June 24, 2014.

Chapter 3
Spin Orbit Torque MRAM

3.1 Introduction

The STT (spin-transfer torque) has emerged as a promising memory technology to provide energy efficient, non-volatile, high density memories with low power dissipation and unlimited endurance. In addition, it offers CMOS compatible architectures with high-speed read and write operations. During the initial phase of the development, researchers envisaged the greater potential of the STT based magnetic random access memory (MRAM) to become an alternate solution of the contemporary memory technologies [1]. In the recent past, with the use of in-plane magnetic anisotropy (IMA) based magnetic tunnel junction (MTJ) devices; the STT-MRAM has been commercialized successfully with the 64 Mb storage capacity [2]. However, these memories with higher critical current requirements (100–200 µA) exhibited lower speed (10–20 ns). The higher energy consumption and poor delay performance have made them uncompetitive to the existing SRAM and DRAM technologies [3]. Furthermore, the STT-MRAM poses some limitations with the designing perspectives. The higher switching current densities imply the usage of MTJs with lower resistance-area (*RA*) product below 5 $\Omega\mu m^2$, which can be realized using thinner MgO tunnel barriers. However, higher switching current with the thinner barrier raises the reliability issues and leads to degradation of related MTJ parameters, such as, tunneling magneto-resistance, write margin, and write speed on the time span. Furthermore, due to common read and write path, there is a possibility of false or accidental writing during the read operation which is also known as "read disturb." Additionally, due to same reason, an individual optimization of read or write operation is not feasible. The aforementioned issues make the STT-MRAM unviable to be used at the higher level of cache stack and embedded memory architectures [4].

One of the solutions to the problems is the use of perpendicular magnetic anisotropy (PMA) based MTJ (PMTJ) devices for the STT-MRAM cell. Unlike IMA devices, these PMA devices are scalable, and hence, with the reduced dimensions

© The Author(s) 2017
B.K. Kaushik et al., *Next Generation Spin Torque Memories*,
SpringerBriefs in Applied Sciences and Technology,
DOI 10.1007/978-981-10-2720-8_3

the write current can be decreased significantly by one-tenth of that of the IMA based MTJ (IMTJ) devices [5]. The reduced write current further decrease the severity of the tunnel barrier reliability problems to a certain extent. However, due to STT architecture, the read and write current must flow through the MTJ stack. Therefore, for STT-MRAM, the decoupling of the read and write paths is not possible, and hence, the read disturb and optimization problems still persist [6].

To mitigate all the aforementioned problems of the STT architecture, an alternative solution known as spin-orbit torque (SOT) technology has been introduced very recently. The SOT structure uses the Rashba and spin-Hall effects (SHE) [7] to flip the free layer magnetization. The SOT devices utilize separate read and write current paths, and thereby, solve the crucial problems of the STT architecture, inherently. Moreover, the SOT-MRAM is very much energy efficient and exhibits faster write access due to independent write path optimization. Since, the SOT-MRAM provides the reliable, energy efficient and fast memory technology solution; it has emerged as a strong contender to replace the SRAM in cache memory [8].

This chapter is organized into seven sections together with this introductory section. Section 3.2 describes the SOT architecture and underlying physics of the device. Section 3.3 discusses the SOT-MRAM bit-cell and array architectures. Section 3.4 explains the SOT-MRAM read and write mechanisms. In the Sect. 3.5, modeling concepts of the SOT-MRAM are illustrated. Section 3.6 discusses the design aspects and optimization of the SOT-MRAM cell. Section 3.7 compares the STT-MRAM and SOT-MRAM memory technologies.

3.2   SOT Device Structure

SOT device structure utilizes a magnetic heavy metal (HM) attached at the top or bottom of the free layer (FL) layer of the MTJ as shown in Fig. 3.1. A strong spin-orbit interaction (SOI) can take place at the FL/HM interface when an in-plane current is passed though the HM. This interaction takes place at the FL/HM interface due to SHE and the Rashba effect. This interaction produces an anti-damping torque due to SHE and a field-like torque produced due to Rashba effect. The strength of the SOI is decided by the magnitude of the current passing through the HM. A current with sufficient strength through the HM can switch the magnetization of the FL of the IMTJ or PMTJ. Furthermore, the SOI depends on the choice of the material of HM, resistivity of the HM, dimensions of the HM and the MTJ, and area of the FL/HM interface. Since, this hyperfine interaction is proportional to the atomic number of the metal, a heavy metal with higher atomic number (Z) is preferable. The most versatile HM materials used in the SOT-MRAM cell are tantalum (Ta), tungsten (W), and platinum (Pt).

The ferromagnetic layer is directed along the width of HM for the proper spin injection. A charge current, I_C is passed through the HM which results in polarization of electrons on the opposite surfaces. The polarization of ferromagnetic layer

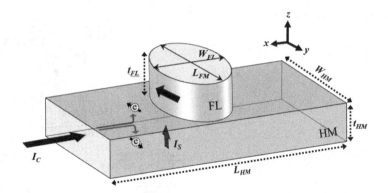

Fig. 3.1 Spin Hall effect

is determined by the charge current, I_C. The scattering of oppositely spin polarized electrons produces the spin current, I_S. The spin current, I_S flows in the z-direction with the spin orientation along the y-direction, if the charge current, I_C flows in the x-direction as depicted in Fig. 3.1. The spin current imparts torque in z-direction. The magnitude of spin current is determined as [9]

$$I_s = P_{SHE}(W_{FM}, t_{HM}, \lambda_{sf}, \theta_{SHE})(\hat{\sigma} \times I_c) \qquad (3.1)$$

where, P_{SHE} is spin polarization, λ_{SF} is spin flip length in Hall metal, θ_{SHE} is spin angle for the HM. The spin angular momentum which is responsible for the spin-torque, can be given by $S = \hbar I_S/2e$ where, $\hbar$ is the reduced Plank's constant and e is the electron charge. If the area of ferromagnetic layer is greater than that of HM, the I_S can be greater than the I_C due to multiple scattering of electrons on the surface so as to impart many units of angular momentum. Hence, in comparison to STT, the SOT produces more torque, however, subjected to the structure dimensions.

3.3 SOT-MRAM Bit-Cell and Array Architectures

The SOT-MRAM bit-cell architecture is shown in Fig. 3.2. The cell is made up of a Hall metal, MTJ, and two access transistors, one for write and other for read operation. Unlike STT device, the SOT device has three terminals, and the SOT based MRAM cell terminals can be connected to the read line, write line, source line, and a word line of the memory array architecture. The word line and source line are connected to the gate and source of the access transistor, respectively. The MTJ is grown vertically over the hall metal. The HM and MTJ perform spin generation and storage operation, respectively. Similar to STT-MRAM, the two resistance states high (bit '1') and low (bit '0') of MTJ are used to store the binary data. The word line is activated to perform the read/write operation.

Fig. 3.2 SOT-MRAM cell
architecture

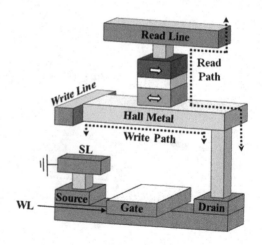

The SOT-MRAM based array architecture is shown in Fig. 3.3 [8]. Every row of cell array consists of data input, output, write circuit, read circuit, and bit cells. The read circuit with a sense amplifier (SA) used to sense the bit information stored as a resistance state of the free layer of MTJ i.e. a low resistance or high resistance representing two distinct logic states.

The use of SOT-MRAM provides some distinct advantages. First, the separate paths for the read and write operations in the SOT-MRAM cell ease the optimization process. Second, the write current that only flows through HM, eliminates the tunnel barrier reliability issues. Third, a lower value of switching current compared to STT-MRAM offers energy efficient solution. However, the promising features of the SOT-MRAM are counter balanced by the requirement of two separate access transistors for the read and write operations and results into increased cell size [7].

3.4 SOT-MRAM Write and Read Mechanisms

The concept of SHE based spin torque is used to write a single bit of information to the SOT-MRAM bit-cell (see Fig. 3.4). Circuit-level schematics of the write and read operations for the SOT MRAM cell are depicted in Fig. 3.4a, b, respectively.

As stated earlier, the SOT-MRAM bit-cell exclusively utilizes two access transistors; one for write operation and the other for the read operation. Furthermore, due to inherent SOI effect and usage of two access devices, separate read and write paths are feasible. As depicted in Fig. 3.4, M_1 transistor is used for write operation, and M_2 transistor is used for read operation. A resistance equivalent model of the SOT-MRAM is presented in Fig. 3.5. As depicted in the figure, the

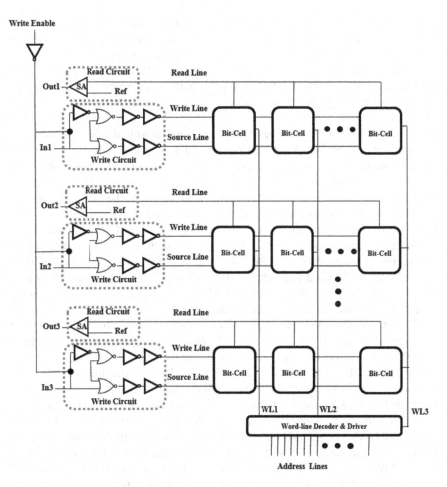

Fig. 3.3 SOT-MRAM cell architecture

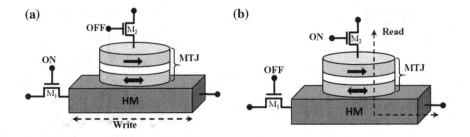

Fig. 3.4 SOT-MRAM **a** write operation **b** read operation

write current (i_{HM}) only flows through the HM and generates SOI depending on the strength of the current and other parameters as described in previous section. Due to very less resistance of the HM compared to the MTJ resistance, the energy required to generate sufficient spin-orbit torque is very less. The current i_{HM} depends on the resistance R_{HM} (= R_{HM1} + R_{HM2}) offered by the HM, which is decided by the resistivity of the HM. The typical value of resistivity of the HM is in the range of 50–200 $\mu\Omega$cm [10].

The read operation is performed in the same manner as in the STT-MRAM cell. However, the magnitude of the read current differs with the inclusion of the Hall metal resistance. The read current i_{MTJ} is decided by the two resistances, R_{MTJ} and R_{HM2}. For the lower RA product of MTJ device, the effect of R_{HM2} cannot be neglected. The inclusion of R_{HM2} lowers the read margin of the SOT-MRAM cell.

The circuit-level implementation of read and write mechanisms is presented in Fig. 3.6 [8]. The transistors T_1, T_2, T_3, and T_4 constitute the write circuit. The transistors T_5 and T_7 are used to select read and write operations. Transistor T_6 along with sense amplifier (SA) form the read circuit. To write the SOT-MRAM, the write select (WR) and write enable (W_s) are activated simultaneously to set the write path and to deactivate the read path. The direction of current flowing between the terminals A and B is decided by the data bit (D_{bit}). To write '0', the transistors T_2-T_3, and transistors T_1-T_4 are made ON and OFF, respectively. The current will flow from terminal B to terminal A to set the polarization of the ferromagnetic layer of SOT-MTJ structure to write '0'. In case of writing '1', the transistors T_1 and T_4 are kept ON while the transistors T_2 and T_3 are made OFF. The direction of current is from terminal A to terminal B. During this operation, the polarization of ferromagnetic layer is set such that the data '1' is written to the SOT-MTJ structure.

To read the data, W_s is made low, which makes the transistor T_6 ON. The transistors T_5 and T_7 are switched OFF and ON, respectively. In addition, the T_1, T_2, T_3, and T_4 are made OFF to dissociate the write circuit from the read operation. The MTJ resistance state is measured with the help of sense amplifier. The pictorial representation of write and read paths are shown in Fig. 3.6.

Fig. 3.5 Equivalent SOT-MRAM resistive network

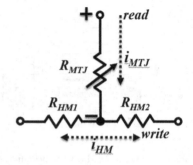

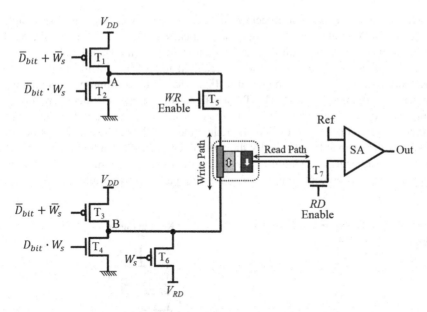

Fig. 3.6 SOT-MRAM write and read circuits in the memory array structure

3.4.1   Concept of Simultaneous Read and Write Operations

During the system level operations, it is required to perform reading from and writing to a memory simultaneously to make the system faster with optimized memory fetch operations. The read-before-write operation mechanism in the modern processing systems employs separate read and then write operations, which leads to slower memory access time. In the aforementioned section, only one operation, either read or write is described by switching ON/OFF the specific access devices. Wherein, the read and write operations were made exclusive. Hence, a read operation on the fly or during the write operation is required to reduce the access time. Due to separate read and write paths in the SOT-MRAM architecture, it is possible to conduct simultaneous read and write operations [11]. This feature utilizes the fact that read operation requires much lesser time than the write operation; hence, concurrent initialization of both the operations result in reading the stored data before the change takes place due to write operation. During the initial phase of the combined operation, reading of the bit-cell performed first and in parallel the write current through the HM start applying torque to flip the magnetization of the FL, and eventually, change of the data, which takes longer time than reading the data.

Unlike the STT-MRAM, SOT-MRAM architecture facilitates the simultaneous read and write operations due to three terminal structure and symmetrical write path. As shown in Fig. 3.7, the write '0' and write '1' path is the same as explained in the previous section. The direction of the write current decides the logic value stored in the MTJ element. However, the read current which flows across the MTJ element, takes the same path that of the write current in the Hall metal.

A modification at the circuit-level is required for performing the simultaneous read and write operations as depicted in Fig. 3.8a. With the appropriate switching of the three access transistors—T_1, T_2, and T_3 ensure the three distinct operations of only-read, only-write, and combined read and write as mentioned in Fig. 3.8b. The timing diagram of Fig. 3.9 depicts the requirement of the read-write access device period which is mainly decided by the write operation. The read and write operations begin after the activation of the access devices. The shorter duration of the read operation is evident from the figure as compared to write duration.

Fig. 3.7 Simultaneous read and write operation in SOT-MRAM cell

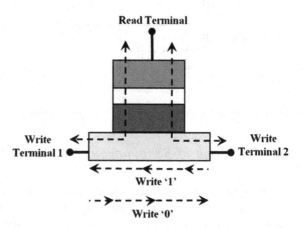

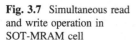

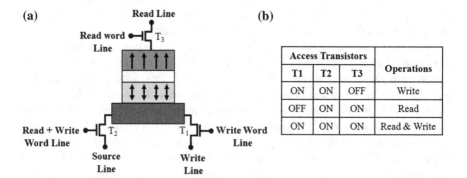

Fig. 3.8 Simultaneous read and write **a** circuit-level schematic **b** access mechanisms

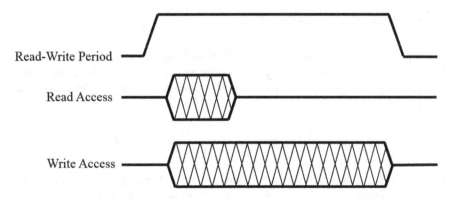

Fig. 3.9 Timing diagram of simultaneous read and write operations

3.5   Compact Modeling of the SOT-MTJ Device

An SOT-MTJ structure plays a key role in the operation of an SOT-MRAM. Therefore, for the accurate design of the SOT-MRAM, the physical behaviour of the SOT-MTJ is required to be investigated [12]. Therefore, the physical modeling of the SOT-MTJ is necessary to trace the read and write operations. The model is composed of magnetization dynamics and *TMR* (tunnel magneto-resistance) blocks. As shown in Fig. 3.10, A, B, and C are the three external terminals of the model. The LLG equation block generates the m_x, m_y, and m_z, the components of the magnetization dynamics of the free layer. These components are given as input to the *TMR* (tunnel magneto-resistance) block. The *TMR* block calculates the tunneling conductance. Depending on the magnitude of the conductance, the bit information stored in the free layer is sensed.

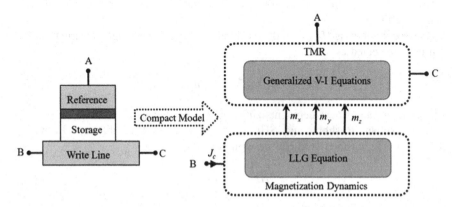

Fig. 3.10 Compact model of SOT-MRAM

3.5.1   Magnetization Dynamics

The input to the LLG module is current density J_c representing the charge current between terminals B and C. The LLG module generates three vector components m_x, m_y, and m_z. The LLG equation representing the magnetization dynamics of the free layer is

$$\frac{\partial \vec{m}}{\partial t} = -\frac{\gamma_0}{1+\alpha^2}(\vec{m} \times \vec{H}_{eff}) - \alpha\frac{\gamma_0}{1+\alpha^2}\vec{m} \times (\vec{m} \times \vec{H}_{eff}) \qquad (3.2)$$

where, m, α, γ_0, and H_{eff} are the magnetization of free layer, Gilbert damping torque, gyromagnetic ratio, and effective magnetic field, respectively.

The first term in the equation represents the precession of the free layer magnetization around the effective field. The second term in the equation signifies the Gilbert damping torque. The Gilbert damping torque is responsible for the magnetization relaxation towards the resultant field. The resultant field is given as

$$H_{eff} = \vec{H}_k + \vec{H}_d + \vec{H}_R + \vec{H}_{SHE} + \vec{H}_{ex} \qquad (3.3)$$

where, H_k, H_d, H_R, H_{SHE}, and H_{ex} are the magneto-crystalline field, demagnetizing field, Rashbha field, SHE field, and external applied field, respectively.

The magnetization of the free layer in a particular direction depends on the interaction of the magnetic moment with crystalline lattice of the free layer. The interaction results in the self alignment of the magnetization without any external field. Therefore, the magneto-crystalline anisotropy field is given as

$$\vec{H}_k = \left(\frac{2K_u}{\mu_0}\right)m_z\vec{z} \qquad (3.4)$$

where, K_u is the uniaxial anisotropy constant, M_s is the magnitude of self magnetization.

The second term is H_d which depends on the shape of the magnetic structure. H_d can be represented as

$$\vec{H}_d = -M_s(n_{xx}\vec{m}_x + n_{yy}\vec{m}_y + n_{zz}\vec{m}_z) \qquad (3.5)$$

where, n_x, n_y, and n_z are demagnetizing tensor coefficients.

The resultant Rashbha and SHE fields make SOT-MTJ different from the STT-MTJ. The rashbha field can be written as

$$\vec{H}_R = -C_R J_c \vec{y} \qquad (3.6)$$

where, C_R is the Rashbha coefficient.

The SHE field is given by

$$\vec{H}_{SHE} = -C_{SHE}J_c\vec{m} \times \vec{y} \qquad (3.7)$$

where, C_{SHE} is SHE coefficient. The values of constants C_R and C_{SHE} depend on the SOT-MTJ stack. The external applied field H_{ex} together with Rashbha and SHE fields performs the magnetization switching of the free layer.

The free layer magnetization dynamics of the MTJ considering different torques acting upon the device can be given with the use of spherical co-ordinates [13] as

$$\frac{1+\alpha^2}{\gamma H_k}\begin{bmatrix} \dfrac{d\theta}{dt} \\ \dfrac{d\phi}{dt} \end{bmatrix} = T_u + T_d + T_{ex} + T_{SOT} \qquad (3.8)$$

where, different torque components are due to crystalline anisotropy (T_u), demagnetization (T_d), external magnetic field (T_{ex}), and spin torque (T_{SOT}). The spin torque (T_{SOT}) is generated due to current flowing through the HM. The strength of T_{SOT} is determined by the magnitude of the spin current I_S, which in turn generated by the charge current I_C flowing through the Hall metal. The direction of the torque decided by the direction of the charge current [7].

$$I_S = \frac{\hbar}{2q}\frac{A_{MTJ}}{A_{HM}}\theta_{SH}\left(1 - \cosh\left(\frac{\lambda_{SF}}{t_{HM}}\right)\right)I_C \qquad (3.9)$$

where, A_{MTJ} is the cross-sectional area of the MTJ, A_{HM} is the cross-sectional area of the Hall metal, θ_{SH} is the spin Hall angle, λ_{SF} is the spin-flip length, and t_{HM} is the thickness of the Hall metal. Here, the ratio of cross-sectional areas of MTJ and HM ($A_R = A_{MTJ}/A_{HM}$) and t_{HM} play the crucial role in designing the SOT based MRAMs.

3.5.2 TMR

The *TMR* module is between terminals A and C. The inputs to the module are m_x, m_y, and m_z. Depending on the magnetization orientation of the respective layers of SOT-MTJ, the *TMR* module deduce the magneto-resistance variations. The tunneling conductance can be obtained as

$$g(V, mx, mz) = \frac{G_{P0}(1 - 2\beta V + 3\delta V^2)}{1 + \left(\frac{1-(m_x \cos \theta_m + m_z \sin \theta_m)}{2}\right)\left(\frac{TMR_0}{1+\frac{V^2}{V_h^2}}\right)} \tag{3.10}$$

where, β and δ are the constants related to the tunnel barrier. G_{P0} is the conductance at 0 V and 0 K. V is the voltage applied between terminals A and C. TMR_0 is the TMR calculated at low bias. The V_h is the potential at which the TMR value becomes half its value at low bias. The θ_m is the angle made by magnetization of the storage layer with the z-axis.

3.6 Design Aspects and Performance Optimization of SOT-MRAM

While designing an SOT-MRAM cell, a prime concern is the reduction in the write current density and write latency. The requirement of reduction in the write current is related to the reduction in the size of the two access transistors. Fortunately, SOT-MRAM structure with separate read and write paths require very less energy for switching the magnetization of the FL. Hence, it is possible to reduce the size of the access device, and hence, the total footprint area.

As apparent from (3.9), for flipping the FL magnetization at lesser energy cost, it is required to increase the ratio of I_S/I_C. This can be achieved by reducing the I_C or increasing the I_S. As mentioned previously, the SOT-MRAM requires a reduced value of I_C due to HM. I_S can be increased with the proper choice of the dimensions of the HM and FL. The relationship of (3.9) reveals the importance of the cross-sectional areas of the FL and HM, A_{MTJ} and A_{HM}, respectively. As shown in Fig. 3.1, $A_{MTJ} = (\pi/4) W_{FL} \cdot L_{FL}$ and $A_{HM} = W_{HM} \cdot t_{HM}$. To increase the I_S, the A_R should be greater than one. This can be achieved using increased W_{FL} and reduced t_{HM}, as the L_{FL} and W_{HM} are of the same lengths. With these dimensions, the spin current I_S can become greater than the charge current I_C, due to multiple scattering of the electrons at the FL-HM interface which impart the torque multiple times to the magnetization of the FL as shown in Fig. 3.11 [14]. Therefore, with the increased value of A_R, the injected electrons get more surface area of interactions or

Fig. 3.11 Spin torque generation with multiple scattering of electrons

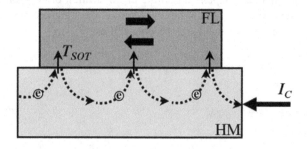

scattering, hence, increased torque. Furthermore, the value of the t_{HM} should be kept comparable to the spin flip length, λ_{SF} in order to make the term $\cosh(\lambda_{SF}/t_{HM})$ negligible for obtaining higher value of I_S.

Due to separate read and write paths, the read and write currents can be optimized independently. The higher magnitude of the write current does not affect the read operation. Similarly, the read action does not affect the writing process, hence, it reduces the possibilities of read-disturb significantly.

3.7   Comparative Analysis of STT-MRAM and SOT-MRAM

The prime difference between the STT and SOT based MRAM cell is the underlying principles used for flipping the FL magnetization in each device. STT-MRAM is a two terminal cell structure with common read and write passage. STT-MRAM due to inherent STT mechanism exhibits asymmetric write operations for storing logic '0' (AP-to-P) and logic '1' (P-to-AP). In fact, the P-to-AP switching of the FL magnetization requires very high energy compared to AP-to-P counterpart. In addition, as the read mechanism utilizes the same path, it is required to design the read current very low in comparison to the lowest write current in order to avoid the read-disturb. However, the reduced read current further decreases the readability of the STT-MRAM cell. Hence, for increasing the readability and the read speed of the cell, it is required to raise the read current, which necessitates the enhancement of the lower threshold of the write current, eventually results into the larger access device, larger footprint area, and higher power dissipation. Therefore, it is utmost required to optimize the STT-MRAM read and write operations with the trade-off between the read and write performances of the cell. STT-MRAM cells with IMTJ devices require the critically optimized write path due to very high switching current necessity; whereas, the cells with PMTJ devices require a delicate balance between the read and write currents due to lower write current threshold. STT-MRAM suffers from the tunnel barrier reliability issues, which can affect the performance of the memory over a span of time. Hence, time dependent degradation of the important parameters is a major concern for the STT based memory architectures and systems.

In contrast to the STT-MRAM, the SOT-MRAM is a three terminal device with two access transistors and one MTJ structure. As mentioned earlier, the SOT-MRAM write operation relies upon the SOI phenomena, exhibits the symmetrical write current for storing logic '0' and logic '1.' Furthermore, the independent read and writes paths offer several advantages such as enhanced tunnel barrier reliability, individual optimization of both the mechanisms, and possibility of symmetric read and write current magnitudes. The symmetric read and write mechanisms require smaller access devices, which results into smaller footprint area. Therefore, despite of the requirement of two access transistors, the

Table 3.1 Comparative analysis of STT-MRAM and SOT-MRAM [6]

Parameters	STT-MRAM	SOT-MRAM
Data storage	Free layer of MTJ	Free layer of MTJ
Non-volatility	Yes	Yes
Area (mm^2)	1.63	1.51
Read latency (ns)	1.2	1.13
Write latency (ns)	11.22	1.36
Read access energy (pJ)	260	247
Write access energy (pJ)	2337	334
Leakage power (mW)	387	254
Process	Hybrid CMOS STT-MTJ	Hybrid CMOS SOT-MTJ
Features	Scalability (✓)	Scalability (✓)
	Endurance (✓)	Endurance (✓)
	Radiation immune (✓)	Radiation immune (✓)
	Retention failure (✗)	Retention failure (✗)

SOT-MRAM exhibits smaller area coverage compared to the STT-MRAM cells. The read access time for the SOT-MRAM is comparable to that of STT-MRAM in the range of 1–2 ns. However, the write latency is significantly reduced nearly to 1 ns, which is one tenth of that of STT-MRAM. Furthermore, the lower switching current requirements and the smaller access device dimensions help in keeping the leakage current to the minimum.

An architecture-level comparative performance analysis of STT-MRAM and SOT-MRAM with size of 512 kB is carried out in [6], which reflects the prospects of the SOT-MRAM bit-cell to be integrated at the higher cache levels. Various performance parameters of the comparison of SOT-MRAM with STT-MRAM bit-cell are listed in Table 3.1.

Conclusively, these attractive features have placed the SOT-MRAM ahead in the race of providing energy efficient, high density, and low power solution required for the next generation non-volatile embedded memory technologies.

Problems

Multiple Choice

1. **In STT-MRAM, read disturb occurs due to**

   a. Separate read and write paths
   b. Common read and write paths
   c. Pinned layer magnetization
   d. Free layer magnetization

2. **The material type used to generate strong Spin orbit interaction is**

   a. Hall metal
   b. Insulator
   c. Intrinsic Semiconductor at room temperature
   d. Lightly doped semiconductor

3. **In SOT-MRAM, the spin current more than charge current is due to**

   a. Low resistance of the Hall metal
   b. Write path optimization
   c. Thinner tunnel barrier
   d. Multiple scattering of electrons on the surface of the Hall metal

4. **The simultaneous read-write operation can be performed with**

   a. Parallel read and write operations
   b. First read operation followed by write operation
   c. First write operation followed by read operation
   d. None of the above

5. **In the following, which is not the advantage of SOT-MRAM**

   a. Energy efficient
   b. Low write voltage requirement
   c. Separate read and write paths
   d. Reduced number of access transistors

6. **In SOT-MRAM, the free layer magnetization switching is due to the**

   a. Torque exerted due to spin polarized current
   b. Torque exerted due to spin current generated by Spin Hall effect
   c. External magnetic field
   d. None of the above

7. **The typical value of resistivity of the HM used in SOT-MRAM is in the range of**

   a. $1–5 \ \Omega m^2$
   b. $10–50 \ \Omega m^2$
   c. $50–200 \ \mu\Omega cm$
   d. $50–200 \ \mu\Omega m$

8. **The effective spin current generation depends on**

   a. Charge current
   b. Spin Hall angle
   c. Cross sectional areas of HM and free layer of MTJ
   d. All of the above

9. **SOT-MRAM is more reliable than STT-MRAM due to**

   a. Separate read and write paths
   b. Utilization of Hall metal
   c. Increased write current
   d. Increased read current

10. **In comparison to STT-MRAM, the footprint area of SOT-MRAM is decreased due to**

    a. Unsymmetrical write current
    b. Size of Hall metal
    c. The size of the access device is reduced due to decrease in write current
    d. Size of the MTJ

Answer Keys: 1-b, 2-a, 3-d, 4-c, 5-d, 6-b, 7-c, 8-d, 9-a, 10-c

References

1. K. L. Wang, J. G. Alzate, and P. K. Amiri, "Low-power non-volatile spintronic memory: STT-RAM and beyond," *J. Phys. D, Appl. Phys.*, vol. 46, no. 7, p. 074003, 2013.
2. www.everspin.com/64mb-spin-torque-mram-ddr3-dram-compatible.
3. H. Yu, Y. Wang. Design exploration of emerging nano-scale non-volatile memory. Springer, 2015, ch. 1.
4. Y. Huai, "Spin-transfer torque MRAM (STT-MRAM): challenges and prospects," *AAPPS Bulletin*, vol. 18, no. 6, pp. 33–40, 2008.
5. S. Ikeda, K. Miura, H. Yamamoto, K. Mizunuma, H. D. Gan, M. Endo, S. Kanai, J. Hayakawa, F. Matsukura, and H. Ohno, "A perpendicular-anisotropy CoFeB–MgO magnetic tunnel junction," *Nat. Mat.*, vol. 9, pp. 721–724, Jul. 2010.
6. R. Bishnoi, M. Ebrahimi, F. Oboril, and M. B. Tahoori, "Architectural aspects in design and analysis of SOT-based memories," *IEEE Proc. Asia South Pac. Des. Autom. Conf. ASP-DAC*, Singapore, pp. 700–707, 2014.
7. Y. Kim, S. Member, X. Fong, K. Kwon, M. Chen, and K. Roy, "Multilevel spin-orbit torque MRAMs," *IEEE Trans. on Elect. Dev.*, vol. 62, no. 2, pp. 561–568, 2015.
8. F. Oboril, R. Bishnoi, M. Ebrahimi, M. Tahoori, G. Di Pendina, K. Jabeur, and G. Prenat, "Spin orbit torque memory for non-volatile microprocessor caches,", *Proc. 1st Int. Work. Emer. Mem. Sol. Conf. DATE, Dresden*, pp. 1–4, 2016.
9. S. Manipatruni, D. E. Nikonov, and I. A. Young, "Voltage and energy-delay performance of giant spin hall effect switching for magnetic memory and logic," *Arxiv*, vol. 103001, pp. 1–16, 2013.
10. G. Prenat, K. Jabeur, P. Vanhauwaert, G. Di Pendina, F. Oboril, R. Bishnoi, M. Ebrahimi, N. Lamard, O. Boulle, K. Garello, J. Langer, B. Ocker, M. C. Cyrille, P. Gambardella, M. Tahoori, and G. Gaudin, "Ultra-fast and high-reliability SOT-MRAM: From cache replacement to normally-off computing," *IEEE Trans. Mul. Compu. Sys.*, vol. 2, no. 1, pp. 49–60, 2016.
11. R. Bishnoi, F. Oboril, and M. B. Tahoori, "Low-power multi-port memory architecture based on spin orbit torque magnetic devices," *Proc. 26th Ed. Gt. Lakes Symp. VLSI*, pp. 409–414, 2016.
12. K. Jabeur, G. Di Pendina, G. Prenat, L. Buda-Prejbeanu, and B. Dieny, "Compact modeling of a magnetic tunnel junction based on spin orbit torque," *IEEE Trans. on Magn.*, vol. 50, no. 99, p. 1, 2014.
13. G. D. Panagopoulos, C. Augustine, and K. Roy, "Physics-based SPICE-compatible compact model for simulating hybrid MTJ/CMOS circuits," *IEEE Trans. on Elect. Dev.*, vol. 60, no. 9, pp. 2808–2814, 2013.
14. X. Fong, Y. Kim, K. Yogendra, D. Fan, A. Sengupta, A. Raghunathan, and K. Roy, "Spin-transfer torque devices for logic and memory: Prospects and perspectives," *IEEE Trans. Compu. Des. Inte. Cir. Sys.*, vol. 35, no. 1, pp. 1–22, 2016.

Chapter 4
Multilevel Cell MRAMs

4.1 Introduction

Over the past three decades, several memory technologies have made their place in the market such as erasable programmable read-only memory (EPROM), electrically erasable programmable read-only memory (EEPROM), static random-access memory (SRAM), dynamic RAM (DRAM), and NAND/NOR flash memories, with varying degrees of commercial success. In general, computer systems employ a memory hierarchy using different types of memories used at different levels. At the highest level, on-chip high speed cache static-RAMs (SRAMs) are used; whereas, at the next higher level, high density, low power off-chip DRAMs are used as a main memory. Flash memories are used to store the large amount of external data owing to its non-volatility features. However, the prevailing memory technologies have entered into the nanoscale regime and encountering various issues regarding their incessant scaling below the 45 nm technology. SRAM exhibits highest speed with access time near to 1 ns; however, lacks in storage capacity and dissipates very high standby leakage power. DRAM faces problems of increasing refresh current and complex physical fabrication process. Flash memories suffer from excess write power, sluggish write speed, reliability, and inadequate endurance issues. In the era of high end mobile computing, it has become imperative to develop embedded systems with high density high speed on-chip memory to match the high processing speed. The aforementioned existing memory technologies are not capable to be used as embedded memory [1].

The spin torque memories have shown tremendous potential to resolve these issues and have laid the foundation to utilize the concept of material-device-circuit co-design to the greater extent. Using the nonvolatile feature of the MTJ devices with switching ability, STT and SOT-MRAMs have shown the potential to implement the concept of "logic in on-chip memory." STT-/SOT-MRAM are non-volatile memories with lower power consumption, high endurance, higher speed, good packing density and scalability suitable for ultramodern memory

© The Author(s) 2017
B.K. Kaushik et al., *Next Generation Spin Torque Memories*,
SpringerBriefs in Applied Sciences and Technology,
DOI 10.1007/978-981-10-2720-8_4

applications [2]. However, till the date, major research was focused on the single level cell (SLC) STT-/SOT-MRAM designs. Furthermore, the SLC configuration of STT-/SOT-MRAM has its own implications and limitations to realize the high density on-chip memory applications [3]. In the recent years, the major thrust is to reduce the cost-per-bit. Hence, the researchers have started to take interest in the multilevel cell (MLC) configuration for the STT-/SOT-MRAMs, which is already a proven technique for the Flash memory technology [4]. Basically, MLC configurations can be obtained by connecting MTJs in series or parallel to represent multiple logic levels. In general, n number of MTJ devices connected in series or parallel represents n-bits or n-levels of logic design. However, the 2-bit MLC STT and SOT-MRAMs in series and parallel configurations with planar NMOS as an access device are discussed in this chapter. The main advantage of MLC structure is the reduction in cost-per-bit as the increment in the requirement of the MTJs does not increase the footprint area of the design, which is primarily acquired by the access device.

This chapter is comprised of seven sections including the current introductory section. Section 4.2 discusses the issues of SLC STT-/SOT-MRAMs. Section 4.3 describes the basic concepts of series and parallel configurations of MTJ structures. Section 4.4 investigates the read and write operations of STT-/SOT-MRAMs using in-plane and perpendicular anisotropy based MLC architectures with series and parallel MTJ configurations. Section 4.5 describes the modeling and simulations of the MLC structures. The designing aspects and optimization of STT and SOT based MLC MRAMs have been put forth in Sect. 4.6. Finally, Sect. 4.7 provides the extracts of the MLC MRAM investigation.

4.2 Issues with Single Level Cell (SLC) STT-/SOT-MRAM

The conventional STT-MRAM offers non-volatility and CMOS process compatibility. However, higher current requirement during the write operation leads to tunnel barrier reliability issues and larger access devices. SOT-MRAM eliminates the reliability issues with strong spin polarized current (100%) and separate read/write paths; however, the additional two access transistors in SOT-MRAM results into increased cell area [5]. The in-plane magnetic anisotropy (IMA) and perpendicular magnetic anisotropy (PMA) based MTJ structures create different impacts on the functionality and performance of the STT and SOT based MRAMs. IMA based MTJ (IMTJ) devices are more suitable for SOT-MRAMs; however, suffers from scalability issues. Contrastingly, CoFeB-MgO material based MTJ with perpendicular magnetic anisotropy (PMTJ) is the promising candidate for sub-40 nm technology nodes [6]. It works well with STT based devices and offers high scalability as the shape anisotropy field-effect is negligible. Despite of all the advantages, the single-level cells (SLC) STT-/SOT-MRAMs are unable to provide the best solution for the on-chip high-density memory applications. One key solution is the multi-level cell (MLC) configuration, a mature technique already utilized successfully in FLASH memories. Multilevel cell (MLC) structure in series or parallel configuration paves a way to circumvent the problems related to the conventional STT-/SOT-MRAM based MTJ devices and provides an enhanced integration density at reduced cost per bit. Conventional STT-/SOT-MRAM requires a unit cell area of ~ 10–$60\ F^2$ and reported simulations have been carried out based on available single-level cell MTJ compact models [7]. However, till date no compact model exists that can capture the device physics of MLC-MTJs accurately. Conclusively, omni-directional (material-device-circuit level) and aggressive research efforts for the development of STT-/SOT-MRAMs are required to achieve the desired performance and characteristics near to that of existing embedded SRAMs and DRAMs.

4.3   Multilevel Cell (MLC) Configurations

Spin torque MRAMs can be configured at multilevel by connecting IMTJ/PMTJs in series or parallel, accordingly they are called as series-MLC (sMLC) and parallel (pMLC). Furthermore, depending on the requirement, the storage devices such as STT or SOT based MTJs can be configured to realize the MLC STT-/SOT-MRAMs.

4.3.1   STT Based MLC Configurations

Figure 4.1 represents an STT based 2-bit sMLC structure using two IMTJs. These IMTJs can be replaced with the PMTJs. Two MTJs with different sizes can be connected in series with a sandwiched non-magnetic (NM) metallic spacer in between to realize the sMLC structure.

Fig. 4.1 STT based 2-bit
series stacked MLC structure

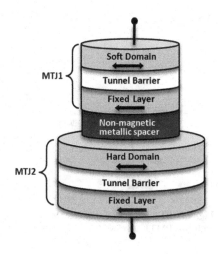

An MTJ with smaller dimensions (hence, low energy barrier, E_b) requires a low switching current for flipping the free layer magnetization, normally called "soft domain." Conversely, the other MTJ with the greater size requires higher current for switching the free layer magnetization, called "hard domain." Comparatively, the realization of sMLC structure is easier than the pMLC structure.

A pMLC configuration can be realized using composite MTJ structure in which two separate free layer domains, normally known as "hard" and "soft" domains, sharing a common insulating barrier and fixed layer as shown in Fig. 4.2a. Due to different coercive forces, magnetizations of both the domains are switched at the different level of current passing through the MTJ. The hard domain with higher coercive force require more current for magnetization switching; whereas, the magnetization of the soft domain is switched at the lower current level. The total free layer area of the composite pMLC structure is the same as of the tunnel barrier and the fixed layer. However, the hard domain width required to have larger dimension along the long axis compared to the soft domain so as to exhibit larger demagnetization field. Henceforth, the composite pMLC structure can only be realized using the IMTJ devices, in which, the different dimensions of the domains require different amount of critical switching current to overcome different demagnetization field. However, the exact reproduction of the desired specifications of the composite pMLC structure is difficult due to exchange interactions of the magnetizations of the two different domains fabricated in the same plane [8].

To reduce the complexity of the process design, another simplistic way of implementation of the pMLC structure is to connect two MTJ devices in parallel with an appropriate distance between each device to justify negligible exchange interactions between the ferromagnets as depicted in Fig. 4.2b. Similar to the sMLC design, a pMLC can be configured using two parallel MTJs with different dimensions. An MTJ with smaller dimension and low switching current represents a "soft bit," and the other MTJ with larger dimension and greater switching current

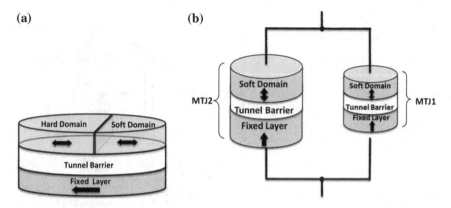

Fig. 4.2 STT based **a** composite MTJ structure for parallel MLC operation **b** pMLC structure with two PMTJ devices

represents a "hard bit." Compared to the composite pMLC structure, this configuration offers a freedom of designing the MRAM cell with individual optimization
of the MTJ devices without sacrificing the footprint area. Furthermore, with this
pMLC configuration, designers have the choice of selecting IMTJs or PMTJs for
the MRAM cell realization.

4.3.2 SOT Based MLC Configurations

SOT based MLC structures can be constructed in similar manner to the aforementioned STT based MLC structures. SOT based sMLC configuration can be
implemented with the addition of a Hall metal to the series-stacked MTJs at the top
or bottom of the free layer of one of the MTJs as shown in Fig. 4.3a. In SOT based
sMLC structures, the spin-orbit interaction resulted from the current passing
through the Hall metal is used to flip the magnetization of only one free layer which
is adjacent to the metal. The dimensions of both the MTJs are different similar to the
STT based sMLC structure.

SOT based pMLC structure using two individual IMTJs connected in parallel
with the two different sizes of Hall metals is shown in Fig. 4.3b. For setting up the
different levels of switching currents in the pMLC configurations, the cross section
areas of the Hall metals and the sizes of the IMTJs are designed at different values.
In this chapter, an in-plane magnetic anisotropy based MTJs are utilized to describe
the SOT based pMLC operations considering the fact that the PMTJ based SOT
structures are still under investigation. Furthermore, due to the possible optimization of the individual MTJ, a structure with two different MTJs connected in parallel
is preferred over the composite MTJ based pMLC structure [5].

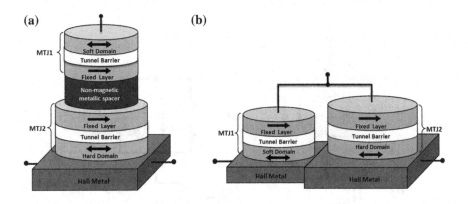

Fig. 4.3 SOT based **a** 2-bit sMLC structure **b** 2-bit pMLC structure

4.4 Multilevel Cell (MLC) MRAM Operations

Similar to the single level cell spin torque MRAMs, the MLC configurations of
STT-MRAM and SOT-MRAM require one and two access transistors, respectively.
The write operation mechanisms in multilevel cell MRAMs depend on the series or
parallel configurations and the type of spin torque devices used i.e. STT/SOT.
However, the read operation is same in all the configurations. In the STT-/SOT-
MRAMs the MTJ devices are fabricated at the back-end-of the line (BEOL) process,
hence, the series stacking and parallel connection of the MTJ devices require comparatively easier process steps and consume very less area compared to the nMOS
devices fabricated at the front-end-of the line (FEOL) process steps [9].

4.4.1 MLC STT-MRAM Write and Read Operations

Series and parallel connected 2-bit MLC STT-MRAM cells are shown in Fig. 4.4a, b,
respectively. The MTJ1 with smaller size is designed to switch at lower current (I_{C1}),
represents a "soft bit." Conversely, the MTJ2 with greater dimensions requires a larger
amount of current (I_{C2}) to switch its magnetization, represents a "hard bit." In this
chapter, the hard bit is considered as a most significant bit and soft bit is considered as
a least significant bit of the 2-bit number. As mentioned in the previous chapters, a
logic level is represented by the distinct value of resistance of a particular MTJ device.
In the current chapter, four distinct values of resistances are denoted as R_{00}, R_{01}, R_{10},
and R_{11} to represent the corresponding four logic levels. Figure 4.5 depicts the
experimental results of resistance-current (R-I) curve for the composite pMLC
STT-MRAM [10] highlighting four ranges of resistance for discriminating four distinct logic levels ("00", "01", "10" and "11"). Furthermore, for each logic level

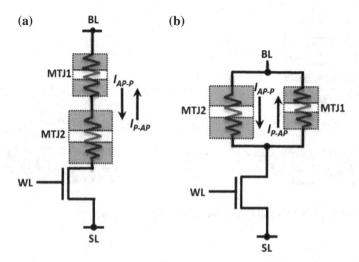

Fig. 4.4 Schematic diagram of STT based 2-bit **a** sMLC MRAM **b** pMLC MRAM

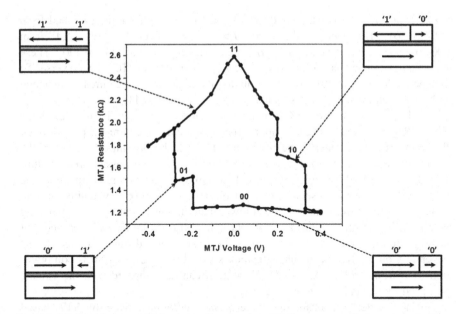

Fig. 4.5 Resistance-current (*R-I*) curve of composite pMLC structure with the magnetization direction of the FL and PL for all four logic states

corresponding relative directions of both the free layers are mentioned in the Figure. As shown in Fig. 4.4, for P-to-AP switching, the bit-line (BL) and word-line (WL) are connected to V_{dd}, and source line (SL) is grounded. The P-to-AP switching current ($I_{p\text{-}AP}$) is considered as a negative current, and conversely, AP-to-P current is considered as positive in Fig. 4.4. For changing the logic from "00" (total resistance around 1200 Ω) to "01," minimum current of -185 µA is to be fed from SL to BL. Similarly, for changing the logic from "11" to "10," a current of 200 µA is to be passed from BL to SL. A direct switching of "10" to "01" or vice versa is not possible.

To understand the mechanism to store a specific logic level among the possible four digital values (00, 01, 10, and 11), a stored logic of "00" is assumed, initially. Figure 4.6 present the switching of the MRAM logic levels depending on the direction and magnitude of the current through the MTJs. Dual direction of an arrow represents current in both the directions for the magnetization switching from

Fig. 4.6 Writing scheme of MLC structures

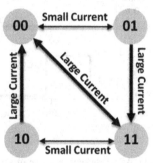

P to AP or AP to P. For the condition, $I_{C1} < I < I_{C2}$, logic can be switched from "00" to "01." However, for the current $I > I_{C2}$, logic level changes to "11" from "00" or "01." A reverse current $I > I_{C2}$ can change the logic of "11" to "00." These logic values can be changed to any level except from "01" to "10" or vice versa. To change the logic from "01" to "10," initially, the logic level is required to switch at "11" and then to "01." This is called two level switching [11].

Read operations in the MLC structures are carried out in similar manner as the SLC MRAMs. A read current of appropriate magnitude is required to pass through the MTJ pillars. In general, the read current should be one third or 20% of the critical current required for the write operation to avoid the possible read disturb [12]. For 2-bit MLC structures of Fig. 4.4, designing a read current is critical as one of the MTJ devices representing the soft bit, requires a lowest switching current. Hence, the read current must be much lower than the lowest value of the switching current for the configuration. For both the 2-bit MLC STT-MRAMs, read current passing through the MTJ nano-pillars generates four distinct voltage levels depending on the free layer magnetization states. In sMLC and pMLC configurations, the whole voltage across the MTJ stack can be compared with the voltage produced by a standard reference generator.

For the better understanding of read and write operations of STT based MLC MRAM, an sMLC structure connected with planar nMOS transistor as an access

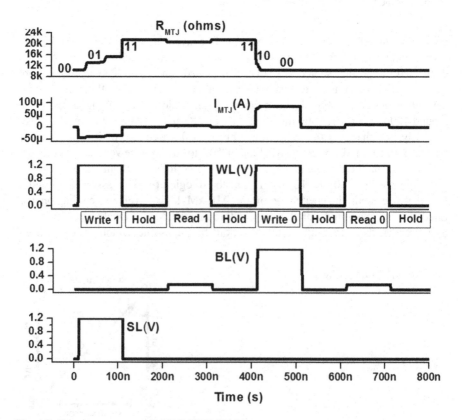

Fig. 4.7 Transient response of sMLC STT-MRAM

device at 65 nm technology node similar to the schematic represented in Fig. 4.4a is simulated and the transient simulations of the design are plotted in Fig. 4.7.

For the simulation, the word line (WL) is connected to a square pulse train with 100 ns pulse width. The bit line (BL) and source line (SL) are excited with the pulse of the same width at different interval of time depending on requirement of the read and write operations. For the write operations, a pulse of 1.2 V is applied to the BL or SL line. To write '1,' SL is connected to 1.2 V and BL is grounded. Conversely, for writing '0' to the cell, BL is connected to 1.2 V and SL is grounded. As the read operation requires current to be fed only one direction, for simulation of the same BL is connected to a pulse of 0.2 V and SL is grounded. When WL is grounded, the access transistor remains in an OFF state leaving the cell in a "hold" state. As can be seen in Fig. 4.7, four resistance states can be achieved with the design. However, the plot reveal the asymmetric nature of the MTJs; switching from "00" to "11" require current pulse with longer duration, while "11" to "00" switching is accomplished in a short duration. This is the reason, the intermediate state "01" can be recognized, but it is difficult to recognize "10" state on the time span. Furthermore, it is evident from the simulation results, during the read operation, the resistance of the MTJ stack deviates very little due to very low voltage requirement for the operation.

4.4.2 MLC SOT-MRAM Write and Read Operation

Figure 4.8 presents the schematics of SOT based 2-bit sMLC and pMLC MRAMs, respectively. As mentioned in the Chap. 3, SOT-MRAM offers a separate read and write paths. However, this is partially correct statement for the sMLC SOT-MRAM structure. Write current flows through the Hall metal (HM) with grounded source line (SL) and bit line and word line (WL$_b$) at V_{dd}. As shown in Fig. 4.8a, for SOT sMLC, free layer magnetization of the MTJ2 only can be switched through the strong SOI effect. However, to flip the logic level of MTJ1, a current $I > I_{C1}$ is required to pass through the MTJ structures. In pMLC structures shown in Fig. 4.8b, write current flows only through the Hall metal, and the change of the logic states in the parallel connected MTJs depends on the strength and direction of the current. The switching mechanism follows the same steps described in Fig. 4.6. As evident from Fig. 4.8a, sMLC is less energy efficient than the pMLC structure as the STT based switching of MTJ1 require more current compared to SOT based switching current. However, in general, SOT based MLC structures require less switching current, and hence, more energy efficient than the STT based MLC configurations.

The read operation is similar to the STT based MLC MRAM configurations. However, the design of read current parameter for pMLC is rather critical than the sMLC structure. In sMLC configuration the read current generates a voltage across the resistance of series stacked MTJs. Depending on the lowest level of series resistance the read current is designed. However, in pMLC structure, the read current is divided into the parallel MTJ branches, and hence, require more careful design of the same to differentiate between the successive logic levels.

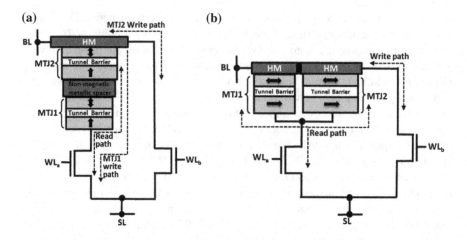

Fig. 4.8 Schematic diagram of 2-bit SOT-MRAM **a** sMLC **b** pMLC configuration

4.5 Modeling and Simulation of MLC MRAMs

An ingenuous analysis of STT/SOT based MLC designs requires an accurate MTJ model imitating the effect of anisotropy fields, thermal fluctuations, and spin torque dynamics. Several SPICE compatible models for an MTJ device have been suggested in the recent past, which can utilized to replicate the behaviour of MLC structures [13–18]. A SPICE compatible simulation framework [19] is depicted in Fig. 4.9, which is imitating one MTJ. The complete model is divided in two parts. The first part depicts the LLG module which accepts the inputs from various fields, including the thermal fluctuations module, to be utilized by the well-known Landau-Liftshitz-Gilbert-Slonczewski (LLGS) formalism. This part presents an application of the LLGS formulation and modeling of the other parametric relationships including the thermal fluctuations. Finally, the third stage accepts the positional values of the magnetization vectors, which eventually decides the resistance of the MTJ.

The free layer magnetization dynamics of the MTJ considering different torques acting upon the device with the use of spherical co-ordinates can be given by the following relationship [20]

$$\frac{1+\alpha^2}{\gamma H_k}\begin{bmatrix} \dfrac{d\theta}{dt} \\[2mm] \dfrac{d\phi}{dt} \end{bmatrix} = T_u + T_d + T_{ex} + T_{STT/SOT} \qquad (4.1)$$

where, different torque components are due to crystalline anisotropy (T_u), demagnetization (T_d), external magnetic field (T_{ex}), and due to spin torque $(T_{STT/SOT})$, H_K is the anisotropy field, α is the damping factor, and γ is the gyromagnetic ratio.

Fig. 4.9 Simulation framework for MLC architecture

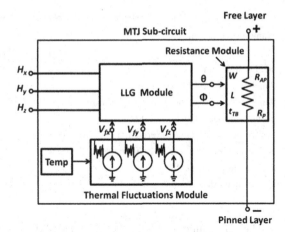

The spin torque ($T_{STT/SOT}$) is generated due to current flowing through the MTJ structure in STT devices, and through HM in SOT devices. The strength of T_{SOT} is determined by the magnitude of the spin current (I_S) injected by the charge current (I_q) flows through Hall metal (HM) and the direction of the torque decided by the direction of the current [21]. The spin current, I_S can be expressed as

$$I_S = \frac{\hbar}{2q}\frac{A_{MTJ}}{A_{HM}}\theta_{SH}\left(1 - \cosh\left(\frac{\lambda_{SF}}{t_{HM}}\right)\right)I_q \tag{4.2}$$

where, A_{MTJ} is the cross-sectional area of the MTJ, A_{HM} is the area of the Hall metal, θ_{SH} is the spin Hall angle, λ_{SF} is the spin-flip length, and t_{HM} is the thickness of the Hall metal. Here, the ratio of cross-sectional areas of MTJ and HM ($A_R = A_{MTJ}/A_{HM}$) and t_{HM} play the crucial role in designing the SOT based sMLC and pMLC MRAMs.

4.5.1   Simulations of MLC MRAMs

In this section, simulations of the sMLC and pMLC STT-MRAMs are presented using the framework depicted in the previous section. Moreover, the impact of IMTJ and PMTJ devices on the performance of the MLC structures is discriminated using the simulations results. Herein, the simulations are performed with the assumption of negligible exchange interactions between the ferromagnetic layers. The performance analysis of the SOT based MLC structures can be carried out in the similar manner considering the modeling relationship described in (4.1) and (4.2). Physical dimensions and other parameters of the IMTJ and PMTJ devices used for the simulations are listed in Table 4.1.

Table 4.1 Physical and electromagnetic parameters of MTJ devices used for sMLC design

Name of the parameter	IMTJ [22]		PMTJ [23]	
	MTJ1	MTJ2	MTJ1	MTJ2
FL/PL area (nm^2)	140×70	150×75	50×50	60×60
Thickness of the FL (t_{FL}) (nm)	4	5	1.4	1.4
Thickness of the tunnel barrier (t_{TB}) (nm)	0.8	0.8	1.0	1.0
Saturation magnetization (M_S) (emu/cm^2)	800	800	800	800
Anisotropy constant (Ku) (erg/cm^2)	2.6×10^5	2.6×10^5	4×10^6	4×10^6
Resistance-area product (RA) ($\Omega\mu m^2$)	20.00	20.00	1.00	1.00
TMR at zero bias (TMR_0)	130	130	55	55
Gilbert damping constant	0.01	0.01	0.01	0.01

The simulation result of IMTJ based sMLC is presented in Fig. 4.10. For elliptical shaped IMTJ based sMLC design, MTJ1 with smaller and MTJ2 with larger sizes are used. The RA product is high (20 $\Omega\mu m^2$), due to which, the total antiparallel resistance in "11" logic state achieves the magnitude of ~8 kΩ. The inherent asymmetric nature of the MTJ devices becomes apparent from the simulation results. The soft domain flips at the −200 and 375 μA; whereas, the hard domain flips at the −300 and 500 μA. As can be seen, the sMLC demonstrates a poor resistance margin during the switching from "11" to "01" or vice versa.

The difference in the resistance between the two states is around 200 Ω, which is very small to be distinguished. The higher current is produced with the application of the higher voltage across the MLC stack, which in turn degrades the TMR rapidly, and hence, at the lower end the resistance margin reduces significantly. Thereby, designing of the sMLC STT-MRAM requires critical optimization and trade-off of the parameters.

Fig. 4.10 Resistance-current (R-I) curve for IMTJ based sMLC structure

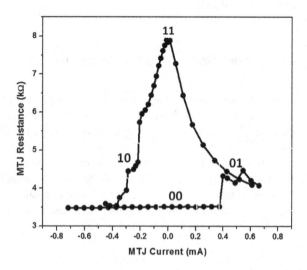

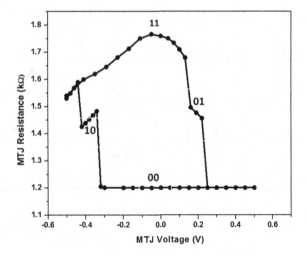

Fig. 4.11 Resistance-voltage (*R-V*) curve for PMTJ based sMLC structure

The cylindrical PMTJ based sMLC design uses the scaled MTJ devices with smaller diameters compared to IMTJ counterpart. The lower *RA* product of 1 $\Omega\mu m^2$ with *TMR* of 55% is chosen. Due to small MTJ device dimensions and lower *RA* product, the switching current requirement reduces, and hence, the MLC stack can flip the state at lower potential as revealed from the simulation results of Fig. 4.11. The lower potential requirement results into lesser *TMR* degradation, which reflects with the higher resistance margin. Hence, PMTJ based sMLC design is energy efficient and provide better resistance margin in comparison to the IMTJ based sMLC design.

Lou et al. demonstrated a single IMTJ based pMLC using two free layers with different coercivities [8]. The pMLC structure offers lower resistance for the obvious reasons. For the simulation, two IMTJs with different coercivities are connected in parallel. In addition, negligible exchange interactions between the nanomagnets are assumed for simplicity.

As listed in Table 4.2, a very high *RA* value of 50 $\mu\Omega m^2$ is chosen, which results into total R_{AP} of 2.6 kΩ at the given dimensions and *TMR*. The lower resistance

Table 4.2 Physical and electromagnetic parameters of MTJ devices used for pMLC design

Name of the parameter	IMTJ [24]	
	MTJ1	MTJ2
FL/PL area (nm^2)	200 × 100	200 × 100
Thickness of the FL (t_{FL}) (nm)	2.0	2.0
Thickness of the tunnel barrier (t_{TB}) (nm)	1.1	1.1
Saturation magnetization (M_S) (emu/cm^2)	800	975
Anisotropy constant (*Ku*) (erg/cm^2)	2.6 × 10^5	2.6 × 10^5
Resistance-area product (*RA*) ($\Omega\mu m^2$)	50.00	50.00
TMR at zero bias (TMR_0)	130	130
Gilbert damping constant	0.01	0.01

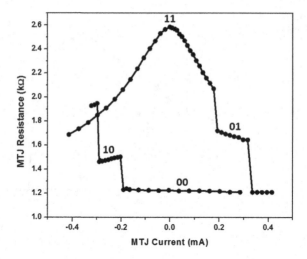

Fig. 4.12 Resistance-current (*R-I*) curve for IMTJ based pMLC structure

signify the requirement of lower potential, and hence, less *TMR* degradation and
higher resistance margin. However, IMTJs require larger switching current; the *TMR*
degradation is larger than the PMTJ counterpart. As evident from the simulation
results of Fig. 4.12, all the four resistance states are very much distinguishable.

4.6   Design Aspects and Optimization of MLC MRAMs

The key issue in designing 2-bit MLC MRAM is to produce four distinct resistance
levels to represent all the logic states in correct and reliable manner with enough
read and write margins. As mentioned before, in order to achieve distinct states of
resistance the switching current of the individual MTJ is to be adjusted at different
values by keeping the sizes of the MTJs different. The correctness of these four
distinct resistance states depends on the physical dimensions of the MTJ1 and
MTJ2. The following are the physical parameters of the MTJ required to be
engineered to set the desired critical current: (1) cross-sectional area of the MTJ
(A_{M1} and A_{M2} for MTJ1 and MTJ2, respectively), (2) tunnel barrier thickness (T_{B1}
and T_{B2}), and (3) free layer thickness of the MTJ (T_{FL1} and T_{FL2}). Depending on the
configuration (sMLC or pMLC), the type of the MRAM (STT or SOT), and ease of
the fabrication process, these parameters are set to different design values. In fact,
the tunnel barrier thickness determines the resistance-area (*RA*) product of the MTJ,
which plays very crucial role in the design and operation of the MLC MRAMs.
However, different tunnel barrier and free layer thickness implementations require
additional fabrication steps, and hence, the technique of realizing different
cross-sectional area of the MTJs are preferred.

4.6.1   sMLC MRAMs

For sMLC STT-MRAM designs, the series-stacked MTJs can be designed with different cross-sectional areas as shown in Fig. 4.1. The smaller MTJ1 can be designed with minimum feature size. The larger size of MTJ2 reduces the resistance of the device; however, it is compensated by the requirement of the higher switching current due to volume increment, and hence, requires higher switching energy. The larger dimension of the MTJ does not increase the area requirement of the cell; however, the demands for the larger access device dimension, which reduces the density of the MRAM array. On the other hand, for the case of different tunnel barrier thicknesses (T_{B1} and T_{B2}) at the same cross-sectional area, one of the MTJs is required to be designed with higher thickness, and hence, higher RA value. Eventually, this leads to increase in the overall resistance of the sMLC structure compared to the previous case of different cross-sectional areas. The higher RA value results into higher switching current requirement and poses limitations on the driving capacity of the access transistor and affects the write performance. Furthermore, the realization of different thicknesses of tunnel barrier requires more processing steps than implementation of MTJs with different areas. Hence, MTJs with different cross-sectional areas at the same T_B are more preferable.

The sMLC STT-MRAM use the same read and write paths, hence, while writing the MTJ2 with larger switching current, unintentional MTJ1 writing operation is performed. As a result, wrong data may be stored into MTJ1. This undesired writing is known as "write-disturb failure" [5]. In order to correctly write the MTJs, two-step writing scheme is adapted. Initially, the MTJ2 is written and then MTJ1 is again written with the smaller value of switching current.

The SOT based sMLC designing can be accomplished in similar manner as STT-MRAMs. The cross-sectional area of MTJs and Hall metals (HMs) can be set to different value or MTJs with different values of TB can be fabricated. As described in the Chap. 3, the spin current (I_S) generated by the charge current (I_q) flowing through the HM, is directly proportional to the cross-sectional area of the MTJ. Hence, wider MTJ produces larger I_S, and hence, flips at lower value of current. This leads to increase the write-error rate (*WER*). This issue can be addressed by the use of spin sink layer (SSL) attached to the unused surface of the HM [4]. The SSL increases the spin current without the need of wider MTJ device. Hence, with the use of SSL in one of the MTJs, the distinction in switching currents can be achieved without the different cross-sectional area requirement.

The SOT sMLC requires very less switching energy (less than 50% of the STT based sMLC) due to different write paths and SOT operation. As depicted in Fig. 4.12, to write the MTJ2, a current is fed through the HM without disturbing the MTJ1, which completely wipe out the possibility of write-disturb failure. Hence, two-step writing operation is not required in this configuration. Due to SOI effect, the MTJ2 can be switched at very less energy compared to the STT counterpart. Furthermore, a smaller current is required to write the MTJ1 representing a soft bit. Thus, the SOT sMLC structure is much energy efficient with low write-error rate (*WER*). However, due to shared read and write path of MTJ1, it is difficult to optimize the performance of these operations separately.

The read operations of STT and SOT based sMLC MRAMs are identical. A carefully designed read current is to be passed through the MTJ stack. However, the requirement of the read current in SOT based sMLC design is less as compared to its STT counterpart. Generally, the read margin (*RM*) of the sMLC design is defined as the minimum of the difference in two resistance states, i.e. $RM = \min$ $(R_{10}\text{-}R_{00}, R_{11}\text{-}R_{01}, R_{10}\text{-}R_{01})$. Hence, for better *RM*, higher *RA* values are required. Furthermore, the tunnel magnetoresistance (*TMR*) plays very crucial role in the determination of four distinct resistance states. *TMR* is voltage dependent parameter and determine the separation between the high and low resistance state of the MTJ. Hence, a very high value of *TMR* is desirable. The current MTJ devices exhibit the *TMR* in the range of 80–150%. However, due to series resistance in sMLC, higher voltage is to be applied for generating appropriate read current, which in turn, decrease the *TMR* of the MTJ devices and affect the *RM*.

Conclusively, sMLC designs require higher *TMR*, *RA* product, and A_R in order to achieve higher *WM* and *RM* for correct representation of different logic levels. However, higher A_R requires larger I_C, and higher *RA* product requires higher potential to generate requisite I_C. This eventually, leads to degradation of *TMR* drastically, and finally, less *RM*. Hence, all the aforementioned parameters are required to optimize for correct sMLC operations.

4.6.2 pMLC MRAMs

In STT based pMLC designs, the read and write operations are identical to the sMLC designs due to common operational path. However, the total resistance of the parallel MTJ structure is lesser compared to sMLC designs. Hence, lower potential is required to generate the switching currents. In order to engineer different switching currents for the parallel MTJ devices, similar techniques described in the previous section can be applied. The write operation can be performed by applying suitable voltage across the parallel formation. Due to parallel configuration, the applied current divides into two parts and the values depend on the present states of the MTJs. After switching of MTJ1 state at lower level of current, the switching of MTJ2 state requires very high current, and hence, potential across the MTJ stack is required to increase substantially compared to first step. The higher current unnecessarily stresses the barrier of the MTJ1, and hence, raises the reliability issues of the MTJ devices.

The SOT based pMLC designs possess a distinct advantage of completely separate read and write paths. This allows the designers to optimize the read and write operations separately. Both the MTJs have the common write current path. Hence, two step writing mechanism is to be followed for avoiding the write-disturb failure. At first step, higher current is passed through the HM to write the MTJ2, which will write the MTJ1 undesirably. In the second step, lower current is passed through the HM to correctly store the logic in MTJ1. However, SOT pMLC designs are very much energy efficient and require 10% of the energy demand of the STT based MLC designs. The prime reason for the energy efficiency is the very light load offered by the HM for writing both the MTJs. Hence, less potential is required to generate higher switching current.

The readout mechanisms for STT and SOT based pMLC designs are identical to that described in the previous section. The overall resistance of the pMLC stack is very less compared to the sMLC designs. Hence, with lower potential the read current can be inserted to the MTJ branches. The lower potential is very much beneficial in order to keep stable *TMR* while read operation is performed. Hence, with the very less *TMR* degradation, distinction between the consecutive resistance states increases, which results in improved *RM*.

The parallel configuration offers another advantage of the selection of higher *RA* product of the MTJ devices. In fact, the parallel configuration necessitates the requirement of higher *RA* product. Higher *RA* product increases the resistance of an individual MTJ; however, the overall resistance of pMLC stack does not increase much due to parallel mechanism. Higher *RA* product results into higher *RM*. Furthermore, the higher *RA* product is achieved by increasing the thickness of the MgO barrier. The increased barrier thickness decreases the reliability issues with pMLC structures. Another advantage of the increased barrier thickness is the reduction in the leakage current. Moreover, the higher *RA* product diminishes the effect of HM resistance. With the *RA* product higher than 10 $\Omega\mu m^2$, the HM resistance can be neglected. For better *RM*, pMLC requires the *RA* product in the range of 25 $\Omega\mu m^2$.

4.7 Conclusions

The simulation results of IMTJ and PMTJ based sMLC and pMLC structures reveal certain important facts while designing the MLC STT-MRAMs. The sMLC structure provides good resistance margins between the logic states. In addition, it works well for low *RA* values, and suitable for more than 2-bit MLC designs. However, series-stacked MLC structure requires a higher value of switching voltage, hence, resulting in less energy efficient design.

On the contrary, as the pMLC structures inherently provide low resistance design, hence, low voltage switching becomes possible. The pMLC designs with lower *RA* values result into the narrow resistance margins between the consecutive logic states, which increases the possibility of false representation of logic levels. However, the pMLC design with higher *RA* product produces promising results. The pMLC structures using PMTJs exhibit four distinct resistance states and extremely low switching voltage at reduced dimensions making them energy efficient and most suitable structures for the IoT applications.

Problems

Multiple Choice

1. **The main advantage of MLC-MRAM is**

   a. Low cost-per-bit
   b. Reduced power dissipation
   c. High spin torque
   d. Minimum standby power

2. **For 2-bit MLC MRAM, two-step writing is required for switching from logic state**

   a. "00" to "11"
   b. "00" to "01"
   c. "01" to "10"
   d. None of the above

3. **Which MLC configuration dissipates minimum switching power**

   a. STT sMLC
   b. STT pMLC
   c. SOT sMLC
   d. SOT pMLC

4. **In the SOT-MRAM, dual write paths are required for**

   a. Series MLC structure
   b. Parallel MLC structure
   c. Series and parallel structures
   d. None of the above

5. **The preferred direction of the read current should be towards**

   a. AP-to-P switching
   b. P-to-AP switching
   c. Either switching
   d. None of the above

6. **In SOT-MRAMs, for generating higher spin current (I_S)**

   a. Ratio of cross-sectional areas of MTJ and Hall metal should be greater than one
   b. Ratio of cross-sectional areas of MTJ and Hall metal should be less than one
   c. Ratio of cross-sectional areas of MTJ and Hall metal should be equal to one
   d. Physical dimensions of the MTJ and HM are to be chosen independently

7. **For a STT-sMLC, MTJs with different cross-sectional areas at the same tunnel barrier width are preferred because of**

   a. Increase in tunnel barrier width increase *RA* product
   b. Decrease in tunnel barrier width increase *RA* product
   c. Increase in tunnel barrier degrades reliability
   d. None of the above

8. **pMLC configuration offers**

   a. Lower potential requirement
   b. Lower *TMR* degradation
   c. Stable resistance states
   d. All of the above

9. **The spin sink layer is used in SOT-sMLC to**

   a. Reduce the read error rate (*RRE*)
   b. Reduce the write-error rate (*WER*)
   c. Reduce the write current
   d. Increase the *RA* product

10. **In STTs-MLC, two-step writing scheme is used to**

    a. Avoid read-disturb failure
    b. Avoid write-disturb failure
    c. To perform the sequential writing
    d. None of the above

Answer Keys: 1-a, 2-c, 3-d, 4-a, 5-b, 6-a, 7-a, 8-d, 9-b, 10-b

Short Answers

1. Explain the issues with single level STT-/SOT-MRAM cell.
2. Enlist the important aspects of STT based sMLC and pMLC.
3. Describe the read/write operation for the SOT based MLC configurations.
4. Derive the important conclusions from the physical and electromagnetic parameters of MTJ devices used for sMLC.
5. Enlist the advantages of SOT based MLCs over STT based MLCs.

References

1. H. Yu, Y. Wang, Design Exploration of Emerging Nano-scale Non-volatile Memory, Springer, 2015, ch. 1.
2. D. Tang and Y. Lee, Magnetic memory fundamentals and technology, 1st ed., Cambridge University Press, 2010, ch. 3–6.
3. Y. Huai, "Spin-transfer torque MRAM (STT-MRAM): challenges and prospects," *AAPPS Bulletin*, vol. 18, no. 6, pp. 33–40, 2008.

4. R. Bishnoi, M. Ebrahimi, F. Oboril, and M. Tahoori, "Architectural aspects in design and analysis of SOT-based memories," *IEEE Proc. Asia South Pac. Des. Autom. Conf. ASP-DAC*, pp. 700–707, 2014.

5. Y. Kim, S. Member, X. Fong, K. Kwon, M. Chen, and K. Roy, "Multilevel spin-orbit torque MRAMs," *IEEE Trans. Elect. Dev.*, vol. 62, no. 2, pp. 561–568, 2015.

6. K. Wang, J. Alzate, and P. Amiri, "Low-power non-volatile spintronic memory: STT-RAM and beyond," *J. Phys. D, Appl. Phys.*, vol. 46, no. 7, p. 074003, 2013.

7. K. Lee, J. Sapan, S. Kang, and E. Fullerton, "Perpendicular magnetization of CoFeB on single-crystal MgO," *J. Appl. Phys.*, vol. 109, no. 12, pp. 123910-1–123910-3, 2011.

8. X. Lou, Z. Gao, D. V. Dimitrov, and M. X. Tang, "Demonstration of multilevel cell spin transfer switching in MgO magnetic tunnel junctions," *Appl. Phys. Lett.*, vol. 93, no. 24, pp. 242502-1–242502-3, 2008.

9. L. Liu, C.-F. Pai, Y. Li, H. W. Tseng, D. C. Ralph, and R. A. Buhrman, "Spin-torque switching with the giant spin Hall effect of tantalum," *Science*, vol. 336, no. 6081, pp. 555–558, 2012.

10. Y. Zhang, L. Zhang, W. Wen, G. Sun, and Y. Chen, "Multi-level cell STT-RAM: Is it realistic or just a dream?" in *Proc. IEEE/ACM Int. Conf. Compu.-Aid. Des. (ICCAD)*, November, 2012, pp. 526–532.

11. Y. Chen, W.-F. Wong, H. Li, Y. Zhang, and W. Wen, "On-chip caches built on multilevel spin-transfer torque RAM cells and its optimizations," *J. Emer. Techn. Compu. Sys.*, vol. 9, no.2, May 2013.

12. Y. Ran, W. Kang, Y. Zhang, J. O. Klein, and W. Zhao, "Read disturbance issue and design techniques for nanoscale STT-MRAM," *J. Sys. Archi.*, vol. 0, pp. 1–10, 2015.

13. L.-B. Faber, W. Zhao, J.-O. Klein, T. Devolder, and C. Chappert, "Dynamic compact model of spin-transfer torque based magnetic tunnel junction (MTJ)," in *Proc. 4th Int. Conf. Des. Tech. Inte. Sys. Nanosc. Era (DTIS)*, Apr. 2009, pp. 130–135.

14. J. Harms, F. Ebrahimi, X. Yao and J. Wang, "SPICE macromodel of spin-torque-transfer-operated magnetic tunnel junctions," *IEEE Trans. Elect. Dev.*, vol. 57, no. 6, pp. 1425–1430, 2010.

15. S. Lee, H. Shin, and D. Kim, "Advanced HSPICE macromodel for magnetic tunnel junction," *Jap. J. App. Phy.*, vol. 44, no. 4B, pp. 2696–2700, 2005.

16. B. Das, and W. Black Jr., "A generalized HSPICE macro-model for pinned spin-dependent-tunneling devices," *IEEE Trans. on Mag.*, vol. 35, no. 5, pp. 2889–2891, 1999.

17. L. Faber, Z. Weisheng, J. Klein, T. Devolder, and C. Chappert, "Dynamic compact model of spin-transfer Torque based Magnetic Tunnel Junction (MTJ)," *4th Int. Conf. on Des. & Tech. of Inte. Sys. in Nano. Era*, Cairo, pp. 130–135, 2009.

18. Y. Zhang, W. Zhao, D. Ravelosona, J. O. Klein, J. V. Kim, and C. Chappert, "A Compact model of perpendicular magnetic anisotropy magnetic tunnel junction," *IEEE Trans. Elect. Dev.*, vol. 59, no. 3, pp. 819–826, 2012.

19. J. Kim, A. Chen, B. Behin-aein, S. Kumar, J. Wang, C. H. Kim, and A. M. Anisotropy, "A technology-agnostic MTJ SPICE model with user- defined dimensions for STT-MRAM scalability studies," *IEEE Cust. Inte. Cir. Conf. (CICC)*, San Jose, 2015, vol. 1, pp. 8–11.

20. G. D. Panagopoulos, C. Augustine, and K. Roy, "Physics-based SPICE-compatible compact model for simulating hybrid MTJ/CMOS circuits," *IEEE Trans. Elect. Dev.*, vol. 60, no. 9, pp. 2808–2814, 2013.

21. M. Kazemi, G. Rowlands, E. Ipek, R. Buhrman, and E. Friedman, "Compact model for spin-orbit magnetic tunnel junctions," *IEEE Trans. Elec. Dev.*, vol. 63, no. 2, pp. 848–855, 2016.

22. G. Panagopoulos, C. Augustine, X. Fong, and K. Roy, "Exploring variability and reliability of multi-level STT-MRAM cells," Dev. Res. Conf. - Conf. Dig. DRC, Texas, USA, June, 2012, pp. 139–140.

23. T. Ishigaki *et al.*, "A multi-level-cell spin-transfer torque memory with series-stacked magnetotunnel junctions," in *Proc. Symp. VLSI Technol. (VLSIT)*, Jun. 2010, pp. 47–48.

24. M. Aoki, H. Noshiro, and K. Tsunoda, "Novel highly scalable multi-level cell for STT-MRAM with stacked perpendicular MTJs," *IEEE Symp. on VLSI Tech.*, June, 2013, pp. 134–135.

Chapter 5
Magnetic Domain Wall Race Track Memory

5.1 Introduction

During the past four decades, semiconductor industry has witnessed a race between the development of processing devices/systems and memory technologies following the Moore's law. With the end of Moore's era on the silicon roadmap, the processing technologies are apparent frontrunner than the memory counterparts in terms of accessing speed and integration volumes [1]. In fact, the failure of prevailing memory technologies to cope up with the speed of the modern processors have pressed the researchers to discover new technologies to meet the requirements of the present and futuristic ultra-high speed communication and data processing applications.

As discussed in Chap. 1, the pyramidal shape of the modern memory hierarchy has become the major bottleneck for the integration of high-speed processors and ultra-high density fast memory technologies on the same silicon chip. At the bottom of the memory hierarchy, the hard-disk drives (HDDs) are with the highest density in terabytes (TB), however, exhibit very sluggish access time in milliseconds and possess the size which is totally unfit for the on-chip integration. To achieve higher speed, existing memory technologies such as SRAMs and DRAMs have to trade-off with the capacity. At the highest level of the memory hierarchy SRAMs exhibit the highest access speed of around 1 ns and capability of on-chip integration with the high-end processors; however, the capacities of these embedded memories are still in the range of kilobytes (KB), which is insufficient for the modern day processors. These high-speed embedded memories are volatile in nature; hence, the retention of the data after power failure or during the OFF mode is the major problem. Furthermore, the contemporary computing system on single board still utilizes different memory technologies, right from HDD to SRAMs, at different architecture levels; which increases the complexity and cost of the systems with lesser efficiency. Hence, a non-volatile memory (NVM) technology with capability of ultra

© The Author(s) 2017
B.K. Kaushik et al., *Next Generation Spin Torque Memories*,
SpringerBriefs in Applied Sciences and Technology,
DOI 10.1007/978-981-10-2720-8_5

high density storage, very low access time, and on-chip integration is the need of the hour for the nanocomputing industry [2].

5.1.1   Limitations of Existing and Emerging Memory Technologies

As the computing industry experiences the pressing requirement of replacing the pyramidal memory hierarchy with a non-volatile on-chip high speed magnetic memory technology, it is appropriate to figure out the constraints of the existing and emerging memory technologies. The hard-disk drives evolved in 1950s, and remained the part of the computing systems as a main memory till today due to its non-volatility and capability of storing huge data required for the current high-end applications. The HDDs offer a least cost-per-bit with the TB storage capacity. However, with the evolution in the data storage capacity, the basic mechanisms of the HDDs remained same. Typically, an HDD device utilizes rotating magnetic drums/sectors with mechanical heads for reading and writing purposes. The current HHDs available in the market have the magnetic drums with rotation speed of 7200 rpm and access time in the range of 10 ms [3]. Most of the processors at the present time operates at the speed of 1 ns; hence, 10 ms of data access time immensely restricts the faster data operations. Due to the mechanical structures used, HDD exhibits very slow read/write operation, and head crashes cause the loss of data and reliability issues. Despite the drawbacks, HDD has survived as primary memory in the market due to its high density data storage, least cost, and non-volatile features.

Due to very long access time and unreliable read/write operation, computing systems utilize the secondary memory technologies based on solid-state semiconductor devices. These types of memory technologies offer high speed of operations and better integration with the current processing architectures. The SRAMs and DRAMs are examples of solid-state memory devices with high performance. These memory technologies with great speed operate at the higher level of the memory hierarchy. However, the speed comes at the cost of data storage capacity. The storage capacity of on-chip SRAMs is in KB and several MB for the external DRAMs. Furthermore, these memories are volatile; hence, used for the manipulation and processing of data rather than the storage. The present computing architectures employ non-volatile solid-state devices known as flash memories. Flash memory provides the cheapest non-volatile solid-state memory option. However, flash memories are slower and unreliable in comparison to SRAMs and DRAMs. The voltage pulse of higher amplitude is required to write data in a flash memory. Furthermore, flash memory exhibits very low endurance of nearly 10,000 cycles. Despite the described problems, flash memory is still a prominent memory technology due to non-volatility, lower cost, and higher storage capacity [4].

Several alternative memory technologies are under investigation, and spintronic memory technologies are among the frontrunners. Spintronic memory technologies such as STT and SOT-MRAMs explained in the earlier chapters are capable of providing solution for the issues pertaining to the memory technologies prevailing in the market. However, STT- and SOT-MRAMs are still not mature technologies that can cover all the levels of the memory hierarchy. The major bottleneck is the size of the access transistor of each MRAM element [5]. A bit-cell of an STT-MRAM utilizes one and SOT-MRAM requires two access devices, as a result, the number of access devices required in an array of these spin-torque memories is very large. Therefore, the footprint area and operational energy required for spin-torque memories are higher. In addition, the logic or circuit level developments using the spin-torque memories are still at the infant stage which can fully realize the concept of "logic-in-memory," reliably. Henceforth, despite the high retentivity, high endurance, and non-volatility features, spin-torque memories in the elemental form still demand much aggressive research efforts to become suitable at all levels of the memory hierarchy. All the other emerging NVM technologies mentioned in Chap. 1 are also not capable of covering the entire memory hierarchy, and hence, forcing the researchers to discover the alternatives very rapidly.

The aforementioned discussion connotes some points to be considered while designing a non-volatile, fast, efficient, and high density on-chip memory technology. The first is the removal of any mechanical moving parts, second is the minimization of the number of access devices, third is the reduction in the read and write operational energy, and fourth is the CMOS process compatibility in the nanoscale regime. The magnetic domain wall (DW) based racetrack memory possesses all the aforementioned features and researchers envisaged it to be well versed at all the levels of memory hierarchy with very high speed read/write operations, huge storage capacity, and compatible with nanoscale CMOS architectures. The DW based memory technology is mainly rely upon the data storage in terms of tiny magnetic domains in magnetic nanowire and sliding the magnetic domains with external energy/field for writing and reading purposes. The DW based racetrack memory (RM) technology has gained the stupendous research interest within short period of time due to possession of all the features required for the universal memory [6, 7].

This chapter contains six sections comprising current introductory section. Section 5.2 describes the fundamentals of domain wall motion in a magnetic nanowire. Section 5.3 explains the operation of DW based elemental MRAM. Section 5.4 comprehensively describes the racetrack memories depicting futuristic investigations required for the establishment of the racetrack memory as a mature memory technology. Section 5.5 describes the logic implementations using the racetrack memories. The final section concludes the chapter.

5.2   Fundamentals of Domain-Wall Motion in Nanowire

In the present scenario, the concept of domain wall motion in a magnetic nanowire has attracted the research community immensely. Initially, in 1978, Luc Berger had predicted the movement of the domain walls in a nanowire using electric current passing through it [8]. The concept was only proved to be correct after the successful implementation of e-beam lithography. Thereafter, the DW motion concept has become very instrumental in developing high speed racetrack memory technology.

5.2.1   Magnetic Domains in Magnetic Nanowire

Magnetization in a magnetic material is separated into regions, and the regions with uniform magnetic orientation are known as magnetic domains. In a steady state condition, each atom in magnetic domain preserves its magnetic orientation in a preferred direction as shown in Fig. 5.1a. Every magnetic domain forms a closed loop from north (N) to south (S) pole. The nearby magnetic domain is aligned always in opposite direction leaving some space between them typically known as domain wall. Each magnetic domain is separated by the opposite magnetic domain. When any external magnetic field is applied or electric current is passed through the material, depending on the orientations of the magnetic domains, the strength of the domains is changed and domain walls tend to move. As shown Fig. 5.1b, an external magnetic field of lower magnitude is able to extend the region of magnetic domain in the same direction of the field. The occupancy of area of the domains is decided by the strength of the applied filed, and sufficiently large field can magnetize all domains in one direction producing a monodomain material as depicted in Fig. 5.1c. In all the discussions presented in previous chapters, ferromagnetic materials with monodomain formation were assumed.

The coverage area and strength of magnetic domains are function of the dimensions of and the field applied to the material. In other words, in the absence of external field, the size of the magnetic domains is decided by the shape anisotropy of the material [9]. By scaling down the dimensions of a magnetic material in a

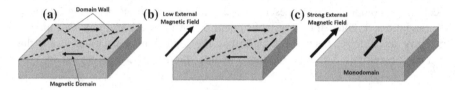

Fig. 5.1 Domain wall presentations in magnetic materials **a** without external magnetic field **b** under low magnetic field **c** under strong magnetic field

nanometer regime, a magnetic nanowire can be formed. Generally, a magnetic nanowire is a very thin and narrow piece of magnetic material containing several domains aligned in longitudinal or perpendicular direction. The more general form of nanowire contains magnetic domains either in-plane or perpendicular to the plane of domain movements. However, due to scalability issues in-plane magnetic anisotropy (IMA) based magnetic materials are less preferred against the perpendicular magnetic anisotropy (PMA) based materials. Figure 5.2 shows a simplified structure of a IMA based magnetic nanowire consisting magnetic domains arranged in opposite directions with respect to the neighbouring domains.

In general, magnetic materials with IMA are greatly affected by the shape anisotropy. When a material width is reduced, the magnetic domains parallel to the width are affected more as evident in Fig. 5.3a. In addition, when the width is few tens of nanometer wide, only longitudinal magnetic domains remain present in the material. As depicted in Fig. 5.2, in-plane magnetic field produces repulsive force at the domain wall edges due to alignment of the similar type of poles (N–N or S–S) at these edges. When the length of magnetic domains is reduced to accommodate more number of bits, the exchange interaction takes place between the nearby domains, eventually reduces the stability of the magnetic material [10].

On the contrary, a nanowire with magnetic domains containing perpendicular magnetic moments less affected by the shape anisotropy and the domains align themselves in the opposite manner as shown in Fig. 5.4a. Due to opposite alignment of the magnetic domains, the magnetic field lines of one domain creates a locking loop with the neighbouring magnetic domain's field lines making the much stable magnetic material. PMA based nanowire can accommodate more number of magnetic domains in comparison to IMA counterpart as it is least affected by the shape anisotropy. In addition, the magnetic domains in PMA materials exhibit higher energy barrier and it requires less energy for the shifting process. Hence, for high density and reliable DW memory the perpendicular magnetization based

Fig. 5.2 In-plane magnetic anisotropy (IMA) based nanowire strip

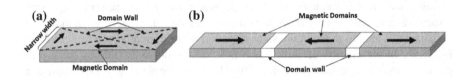

Fig. 5.3 Effect of shape anisotropy on magnetic domains and domain walls **a** material with narrow width **b** material with narrow width and thickness of few atoms

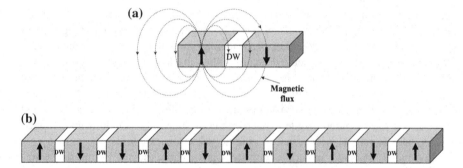

Fig. 5.4 Perpendicular magnetic anisotropy (PMA) based magnetic nanowire

nanowire material is to be chosen. In general, for fabricating magnetic nanowire alloys of cobalt or nickel is preferred [11].

5.2.2   Domain-Wall Motion in Nanowire

As discussed earlier, data access in HDD is based on the mechanical movement of the read/write head laid on the top of the rotational magnetic drums. The entire mechanism is too slow and unreliable. On the contrary, a magnetic nanowire works on the principle of moving the domain walls/magnetic domains using electric current passing through the nanowire. This becomes a biggest advantage since there is no mechanical movement within the structure enabling the device operating at very high speed. Typically, the nanowire operations can be performed in tens of picoseconds, which is much faster than the existing SRAM data access operations (~ 1 ns). The movements of magnetic walls back and forth are essential for the read and write operations from or to the nanowire. The process of creating magnetic domains in the nanowire is in fact the process of storing/writing information on the magnetic material. The straightforward method of generating magnetic domain is to apply a strong local magnetic field in the vicinity of the nanowire. For writing into the next domains, the magnetic domains must be shifted towards right or left using current passing through the nanowire [12, 13].

The mechanism of domain wall movement based on application of current through nanowire is depicted in Fig. 5.5. When an electric current is passing through a magnetic nanowire in horizontal direction, the magnetic domains or domain walls tend to move in the same direction of the current as can be seen from Fig. 5.5b. In the process of shifting the DW, all the domains slide in equal proportion preserving the size of the domains. However, the shifting of DW requires a current density above a certain threshold value to be applied to the end of the nanowire. The magnitude of the threshold current density is decided by the material and shape of the nanowire [14].

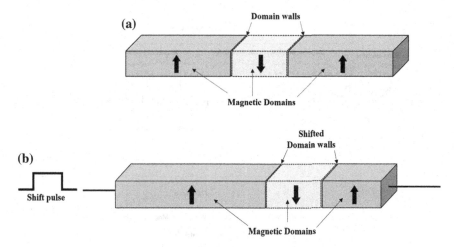

Fig. 5.5 Effect of current pulse on domain wall motion in a magnetic nanowire **a** with zero current amplitude **b** when current pulse is applied

A more detailed shifting operation of current-driven DW in a nanowire is explained using Fig. 5.6. Scattered electrons injected from the left end of the magnetic nanowire are aligned uniformly in the direction of each magnetic domain when they enter into the domain. However, when electrons reach the DW, their magnetizations are progressively bent in the way the local magnetization is bending within the DW. While bending, the angular momentum of electrons is changed; however due to conservation of energy, the reactive force changes the magnetic moments of DW atoms in the opposite direction. This phenomenon generally referred to as spin-transfer torque (STT) effect. Due to this STT effect the magnetization orientation of the domain wall atoms tends to change in the same direction that of the electrons entered, however, pushing the DW further in the direction of the current applied. Conclusively, the current injection moves the DW in the nanowire without changing its size using the STT effect [15]. Therefore, injected current makes the nanowire to work as a shift register [16].

The DW shifting process does not involve any mechanical movement; hence, the energy required to initiate the DW motion is very less. A small amount of DC current above the threshold value is needed to generate the DW motion within the nanowire. However, the DC current generates heat in the nanowire depending on the electrical resistance offered by the material. The Joule heating further increases the resistance of the nanowire resulting into more heat generation. This endless self heating phenomenon can damage the nanowire permanently. Therefore, in practice a voltage pulse of short duration is applied to ensure the excitation is followed by the cooling interval [17].

Theoretically, DW movement should halt when the pulsed current is in cooling period or cut-off; however practically, due to residual thermal energy and reactive components of the nanowire, DW tends to move for a very short period of time after

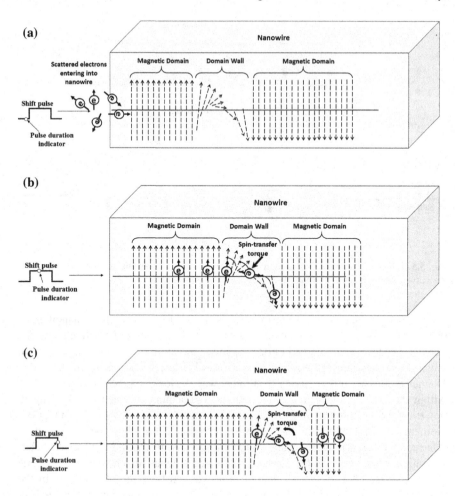

Fig. 5.6 Generation of spin-transfer torque effect using pulsed current for domain wall motion

the pulse is removed. Hence, to stop the DW instantly and manage the self heating in the nanowire, an optimized pulsing mechanism is required to be applied.

5.2.3   Optimization of Domain Wall Motion

To clearly understand all the aspects of DW motion in a nanowire and to optimize it in more controlling manner, further research work is being carried out. Consequently, the researchers are aiming to work in the following areas:

(i) reduction in threshold current density required to instigate DW motion and shifting it, (ii) generating high speed DW motion, and (iii) controlled halt operation of DW motion.

In order to reduce the threshold current density required to start a DW motion in a nanowire from halt position, the nanowire is made thinner and narrower by reducing the width of the domain walls. It is possible to produce very slim nanowire using e-beam lithography. The maximum migration speed of a DW in a nanowire achieved so far is nearly 300 ms^{-1}, which is ten times faster than achieved for HDDs. The investigation and application of different materials are required to build the nanowire for high speed DW motion. Furthermore, the DW velocity (v) is directly proportional to the current injected in the nanowire, and the relationship [18] can be expressed as

$$v = \frac{\beta}{\alpha} \frac{\mu_B P I_{th}}{q M_S} \quad (5.1)$$

where, μ_B is Bohr's magnetron, P is the polarization factor, q is an electronic charge, M_S is the saturation magnetization, β is the nonadiabatic coefficient, and α is the damping constant. Herein, I_{th} is the threshold current to initiate the DW motion from the rest, and it is higher than the current required for sustaining the DW motion in the nanowire. As evident from the equation, the shifting speed can be increased with the use of magnetic materials with higher percentage of polarization. With the fixed amount of I_{th}, magnetic materials with lower value of M_S can increase the DW motion. The value of Ms can be made smaller by decreasing the number of atoms in the magnetic nanowire; however, this option reduces the magnetization stability. Magnetic nanowires made from cobalt based multilayered films demonstrate steady perpendicular magnetization [19].

As depicted in Fig. 5.7, data bits are stored on the track between the small constrictions or traps to overcome the problem of the domain motion due to stray magnetic fields/currents or even due to thermal agitation. These pinned areas are very small and placed at the appropriate distances on the track. Due to small area, a large amount of energy is required to shift the domain or change the magnetization of the domain. Hence, the pinned racetrack structure preserves the data bits against the stray field acting upon the memory and restricts the movement of the domains when no current is applied [20].

Fig. 5.7 Constrictions or trap locations on magnetic nanowire

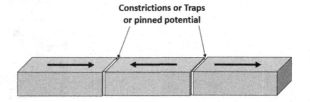

5.3   Domain Wall MRAM

The DW-MRAM device structure consists of two access transistors (only read access transistor is shown in Fig. 5.8), a reference layer, a free layer for the domain wall movement, two anti-parallel pinned layers on the either sides of the ferromagnetic layer, and a tunnel barrier sandwiched between reference layer and free layer to constitute the magnetic tunnel junction (MTJ) [21]. The read line, source line, and word line are connected to the reference layer, source, and gate of the access transistors, respectively.

The width of domain wall depends on type of magnetic anisotropy i.e. IMA and PMA. PMA based magnetic domains have small hard-axis anisotropy and widths can be made narrower due to least shape anisotropy effects compared to IMA materials. Moreover, PMA based magnetic domains have high critical field and low current density. As shown in figure, the domain wall moves between the pinned layers. The movement of domain wall depends on the magnitude of the current from one pinned layer to other to perform the '*bit-write*' operation.

Similar to STT-MRAM, the two resistance states high (bit '1') and low (bit '0') of MTJ are used to store the binary data. The word line is activated to perform the read/write operation.

5.3.1   DW-MRAM Write and Read Operations

The '*bit write*' and '*bit read*' operations are performed with the help of transistors M_1 and M_2 (see Fig. 5.9). To understand the read and write mechanisms, two configurations of each operation are considered in this section. Consider the configurations I and II for write operation, and III and IV for read operation as shown in Fig. 5.9a–d, respectively. For configurations I and II, the transistors M_1 and M_2 are

Fig. 5.8 Simplified architecture of DW-MRAM

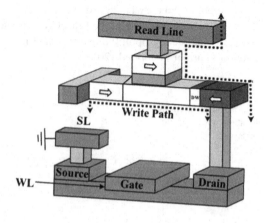

switched ON and OFF, respectively. However, the write currents for these two configurations are complementary to achieve the domain wall motion in opposite direction. The domain wall moves from right to left and left to right for I and II, respectively. The left directed domain wall sets the magnetization of the free layer to the magnetization of right pinned layer to write bit '0'. The free layer magnetization switches to left pinned layer due to right directed domain wall to write '1'.

To understand the read (write) mechanism, the transistors M_1 and M_2 are switched-OFF (ON) and -ON (OFF). The transistor M_2 facilitates the read path between reference layer, tunnel barrier, and only right pinned layer. The low resistance path is offered in III due to parallel magnetization of reference layer and free layer to read the bit '0'. For configuration IV, the reference layer and free layer magnetization are anti-parallel to offer high resistance to read '1'. From aforementioned discussion, it is evident that the '*bit-write*' mechanism depends on the free layer magnetization due to domain wall movement and '*bit-write*' operation is subject to relative orientation of reference layer and free layer.

The DW-MRAM equivalent resistive network is shown in Fig. 5.10. Unlike the SOT-MRAM, the DW-MRAM has low right path resistance due to domain wall structure. For example HM (heavy metal) made up of platinum (Pt) used for SOT-MRAM has low resistance as compared to the Nickel (Ni) for DW-MRAM for the same dimensions [9]. Therefore, the DW-MRAM has low write power dissipation due to low resistive domain wall structure.

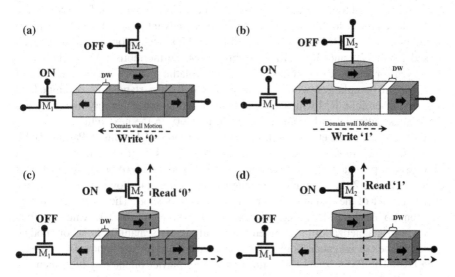

Fig. 5.9 Write and read mechanisms in DW-MRAM **a** write logic '0' **b** write logic '1' **c** read logic '0' **d** read logic '1'

Fig. 5.10 Equivalent
read/write path resistance of a
DW-MRAM

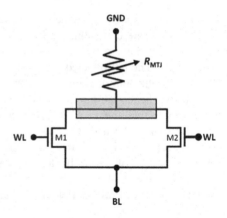

5.4 Racetrack Memory

A nonvolatile memory technology that has low cost-per-bit and high density of an HDD, high read/write speed and reliability of SRAMs or DRAMs, and endurance and retentivity of ST-MRAMs would be a next generation memory technology. Racetrack memory (RM) with all these promising features is suitable for the place [22, 23]. The racetrack name stems from the fact that it contains a track of one or multiple nanowires with information stored in terms of several magnetic domains per nanowire, and these domains are forced to race inside the track for reading and writing data at very high speed. The RM has many advantages as mentioned earlier; in addition to these, RM is much compatible to CMOS technologies and can be well versed for the 2D and 3D memory architectures. Furthermore, RM can be easily employed for the logic circuit designs with nonvolatile memory features. The list of benefits can be enlarged with versatility of RM for the standalone or embedded memory. Therefore, RM which is capable of covering entire memory hierarchy has the greatest potential to become a universal memory in the next future.

The concept of racetrack memory was first proposed by Stuart Parkin (IBM, USA) in 2002. However, the development of RM is still at the infant stage, and only prototypes have been developed so far. The first prototype of RM was demonstrated in 2011 using in-plane magnetic domains [24]; however, due to lower energy barrier of the domains, data retention was very short. Therefore, researchers have employed the PMA based magnetic domain structure in RM, which exhibits very stable magnetization, high retentivity, and greater scalability to accommodate more number of magnetic domains within a nanowire of few hundred nanometer long [10]. The PMA-based RM with high energy barrier offers very high memory density, low power dissipation, and greater access speed.

5.4.1 Structure of Racetrack Memory

A current-induced domain wall based RM consists of track of one or multiple nanowires with read and write mechanisms built using magnetic tunnel junction (MTJ) nanopillars. In general, this whole structure is fabricated at the back-end-of-the-line (BEOL) level on the top of the CMOS architecture; however, for reducing the interconnect delays for the faster access operation, this structure is built close to the CMOS access device. A cross sectional view of a racetrack memory is shown in Fig. 5.11.

As can be seen in the figure, a nanowire strip (racetrack) consists of PMA based magnetic domains. Magnetic domains are separated by the traps or constrictions at appropriate distance to stabilize the magnetizations of the domains against the stray magnetic field or to stop the movement of the domains when shift current is not applied. The racetrack is normally fabricated in the vicinity of the CMOS devices so as to curb the interconnect delays. The read and write heads are placed on the top of the racetrack at the opposite ends or at the specific distance. As shown in figure, MTJ_W and MTJ_R devices are employed to write and read from the racetrack, respectively. The MTJ_W device is wider so as to present low resistance and higher current which flows in bidirectional way using the W_{in} and W_{out} signal lines. At the opposite end, the MTJ_R has the smaller dimension to present higher resistance necessary for the higher sense margins. The read current flows through the R_{in} and R_{out} signal lines. The bidirectional shift current is fed through S_{in} and S_{out} signal path so as to move the magnetic domains back and forth within the nanowire.

The operation of RM is based on the principle of current-induced domain wall movements in a track of a ferromagnetic nanowire. The information/data are stored in the form of magnetic domains, e.g. for PMA based RM, magnetic domain pointed in the upward direction may be considered as logic '1' and downward magnetization as logic '0'. A few hundred nanometer long nanowire track can accommodate several bits of information.

A simplified view of a two-dimensional (2D) planar racetrack memory is depicted in Fig. 5.12. The longitudinal form of racetrack can accommodate lesser number of bits in the available 2D space, hence reducing the density of the RM. On

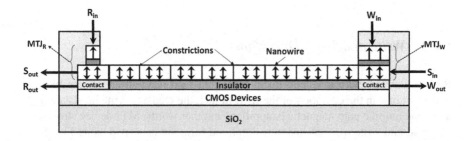

Fig. 5.11 Cross sectional view of a racetrack memory

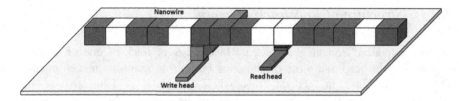

Fig. 5.12 Simplified view of 2-dimensional (2D) planar racetrack memory

Fig. 5.13 Vertical stack of
3D racetrack memory

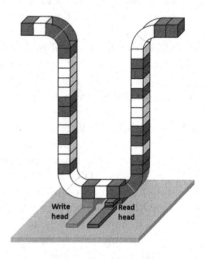

the contrary, magnetic nanowires fabricated in vertical 3D space above the CMOS
structure can store very large amount of data as shown in Fig. 5.13. Furthermore,
3D design space can be used to place an array of racetracks as shown in Fig. 5.14.
Hence, it is possible to fabricate very high density racetrack memory array at the top
of the CMOS devices. With the use of energy efficient PMA devices at the
nanoscale, the 3D architecture is best suitable for on-chip embedded memory
providing very high speed data access. Therefore, RM with vertical 3D nanowire
stack has gained tremendous attraction due to its very high density storage capacity.

5.4.2   Write and Read Operations

Besides the nanowire magnetic track, write and read head structures are playing
important role in the operation of the RM. In fact, the read and write heads are built
using the multilayered magnetic nanopillars, in other words, MTJ device structures.
The read and write mechanisms in a simplified form are presented using Fig. 5.15.
Write and read operations always require the shifting process of magnetic domains
within a nanowire so as to store or sense the data, respectively.

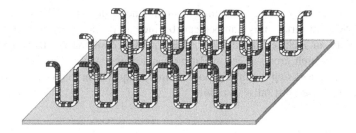

Fig. 5.14 3D array of racetrack memory

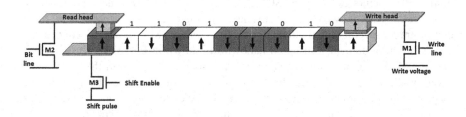

Fig. 5.15 Read and write circuit mechanisms of racetrack memory

For the write operation to be performed, WL is set to '1' to activate the write head (MTJ_w) with larger width. The larger width of the MTJ_w assists the larger current required to nucleate the present magnetic domain underneath the write head. To write into all the bits, sequentially the magnetic domains are shifted from back-to-forth or vice versa. Depending on the data to be stored into magnetic domain, the direction of the current through MTJ_w is changed. The overall write speed depends on the delay in shifting the domain and synchronism to be established between the write and shift operations.

The read operation is performed using the separate read head (MTJ_R) located at the opposite end of the write head as presented in Fig. 5.15. The MTJ_R is designed to have smaller width so as to present higher resistance which is helpful for sensing mechanisms. For reading the bit information present underneath the MTJ_R, BL is set to '1' allowing the relatively smaller current passing through the MTJ_R. Similar to the write operation, the DW shifting process is necessary.

One worth noting fact about RM is that the operation of data access is essentially sequential in nature. In other words, RM is inherently a sequential memory. Hence, it requires more cycles to fetch data from particular location. However, by placing multiple read/write heads data can be fetched more quickly making the RM operations resemble to the random access memory architectures. In general, the average width of a magnetic domain is ~ 100 nm; hence, a nanowire of the length of 1 μm can accommodate ten numbers of magnetic domains. With the use of scalable PMA materials this number can be increased. However, the shifting of domain walls in lockstep decides the accessing speed of the racetrack memory. Recently, a research group of Stuart Parkin has demonstrated the movement of six domain walls in lockstep within a nanowire; this is equivalent to processing 6-bit of information in one shift pulse duration.

5.5   Racetrack Memory Based Logic Implementations

The racetrack memory has brought the paradigm shift in the on-chip logic and memory applications due to its high density as compared to the other emerging nonvolatile memories such as SOT- and STT-MRAMs. Besides the memory application, RM is well versed to develop logic circuits. Researchers have demonstrated several digital circuit applications using RM [18, 25]. A sequence of binary bits in a domain wall based nanowire is shown in Fig. 5.16. The structure has one MTJ at both extremes. The right MTJ is used to perform the write operation, while the left MTJ read the bits move in a sequence. Each bit is stored as magnetic domain oriented in upward (logic '1') or downward (logic '0') direction formed between domain walls. To store the stream of bits, a sequence of write operations is performed on MTJ with the help of train of pulses. As a result, domain wall shifts the data from right to left. The bit reached at the left MTJ is sensed or read when a current is passed through the left MTJ.

The aforementioned domain wall based RM structure can be used to implement digital logic applications. Figure 5.17 depicts an example of generic structure using RM for logic implementation which typically employs multiple excitations with time dependent bit values. As shown in Fig. 5.17, an individual input pattern (A, B, or C) is initially stored using separate RM structure and DW motion using pulsed operation initiated from left to right. The inputs are read by the right MTJs to make them available to perform the desired function at the logic block. The logic block produces the output Y. The output is written with the help of left MTJ of the output free layer nanowire. The output bit is moved to the right MTJ through the domain wall movement. At the right MTJ, the bit information as MTJ resistance state is read by passing the current through the MTJ.

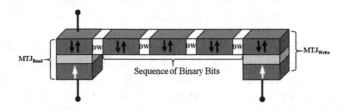

Fig. 5.16 Sequence of bit storage in racetrack memory

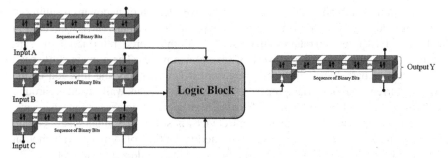

Fig. 5.17 Generic structure of logic implementation using racetrack memory

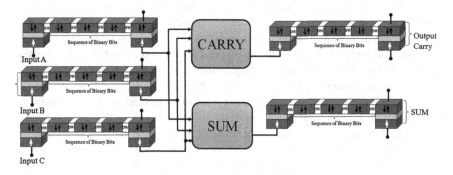

Fig. 5.18 Schematic of RM based multi-bit adder

The racetrack memory has shown the prospects of implementing the concept of logic- in-memory. However, the synchronization of domain wall shift in individual input nanowire is essential. The different domain wall shift velocities in input nanowires may produce false result at the output. Therefore, the current flowing through each of the nanowires is required to be carefully tuned to shift the respective domain walls synchronously to get the correct output.

As a capstone example, a racetrack based full adder is presented in Fig. 5.18 [25]. The inputs A, B, and input carry C are given to the respective write MTJs. The *SUM* and *CARRY* operations as shown in Fig. 5.18 are performed on the inputs to shift the outputs through the respective nanowires towards the read MTJs.

5.6   Chapter Summary

In this chapter, the structure, underlying physics, and applications of domain wall based memories are presented. A domain wall is formed at the boundary of the antiparallel magnetic domains in a nanowire structure. With the less effective shape anisotropy, the PMA based materials have narrow DW compared to IMA; hence, can accommodate higher number of bits in a given length of nanowire resulting in very high density. The domain wall based applications are inherently sequential in nature and require separate mechanism to shift the stored information in synchronous manner among the network of several nanowires. The domain shift speed is the crucial parameter while designing, which is essentially the function of the pulsed current and material property. The typical domain wall speed is around 100 ms^{-1} and needs to be increased for ultra high speed of data access mechanisms. As the development of the domain wall devices is still at the infant stage, certain areas such as reduction of initialization energy, halting of domain wall motion require further optimization exploration. For stopping the domain wall motion at no current condition, the concept of constriction or traps is discussed in the chapter.

The DW MRAM is basic application of the concept of domain wall motion. The DW-MRAM offers several advantages over other MRAM technologies in existence. Racetrack memory is another perfect application of domain wall concept, whereby, the domain wall motion is controlled in a single or multiple nanowires with read and write mechanisms for high speed data access operations. The greatest advantage of the RM is integration compatibility with the CMOS devices in 2D and 3D design space resulting in very high density memory storage. The absence of moving mechanical parts and least energy required to control the movement of DWs make RM best suitable for universal solution covering the entire memory hierarchy. Furthermore, another dimension of RM is the its versatility for the realization of the concept of "logic-in-memory," whereby the stream of data stored on nanowire strips can be accessed for logic manipulation with the help of synchronous network of racetracks comprising several read and write heads typically built using MTJ devices.

Racetrack memories have tremendous potential to be established as a next generation memory and have gained remarkable attention of the research community. However, RM is in the initial phase of the development and still restricted to prototype demonstrations. Being a sequential memory structure, it requires a perfect synchronism of the several domain wall operations confined in multiple racetracks of nanowire network. Different delays of separate nanowire DW shift operations can become an issue for the correct data access or logical manipulations. In order to produce high speed high density memory device, controlled and optimized DW motion is required, which highly depends on the material property and design optimization techniques. The density and cost-per-bit of RM are decided by the number of bits accommodated in a racetrack, and the access speed is determined by the number of domain walls simultaneously moving in a lockstep condition. Recently, the group of Stuart Parkin has demonstrated the movement of as many as six domain walls in a nanowire strip [22]. The PMA based RM memories exhibit high integration capability, scalability, and least operational energy with stable magnetization. However, the PMA based RM faces greater challenges at fabrication level in order to produce nanoscale memory architecture. Keeping in mind the future application requirements, exploration of efficient and fast materials, development of design optimization techniques and realization of relatively easy fabrication process are the biggest challenges for the domain wall based racetrack memories.

Problems

Multiple Choice

1. **In a magnetic nanowire, the velocity of domain wall movement depends on**

   a. Current injected into the nanowire
   b. Size of nanowire
   c. Number of magnetic domains in the nanowire
   d. None the above

2. **A short duration voltage pulse is used to shift the domain wall for**

   a. Increasing the speed of the domain wall
   b. Reducing the domain wall width
   c. Reducing the Joule heating effect
   d. Reducing the number of magnetic domains formed along the nanowire

3. **The trap sites are inserted in a magnetic nanowire to**

   a. Store the data bit
   b. Reduce the Joule heating effect
   c. Reduce the width of the domain wall
   d. Curb the residual movements

4. **In comparison to IMA nanowires, the PMA nanowires have**

   a. Small hard-axis anisotropy
   b. Large hard-axis anisotropy
   c. Low critical field
   d. High current density

5. **For logic applications, the racetrack memory facilitates**

   a. Parallel data bits
   b. Sequential data bits
   c. Multiplexed data bits
   d. None of the above

6. **The Racetrack memory can be made as random access memory by**

   a. Adding the number of read heads at multiple locations along the racetrack memory.
   b. Reducing the size of the nanowire utilized for the racetrack memory
   c. Utilizing the 3D architecture of the racetrack memory.
   d. Reducing the magnitude of the current required to shift the domain wall.

7. **The racetrack memory has**

   a. Low density, sequential data access
   b. High density, sequential data access
   c. Low density, parallel data access
   d. High density, parallel data access

8. **The domain wall width depends on**

   a. Magnetic anisotropy
   b. Size of nanowire
   c. Number of domains formed along the magnetic nanowire
   d. None of the above

9. **The number of atoms along a nanowire is reduced to**

   a. Increase the saturation magnetization
   b. Reduce the saturation magnetization
   c. Reduce the resistance of the nanowire
   d. Reduce the conductance of the nanowire

10. **The magnetic material becomes unstable due to**

    a. Exchange interaction takes place between neighboring domains
    b. Magnetic leakage through the read heads
    c. Magnetic leakage through the write head
    d. Exchange interaction between read and write heads

Answer Keys: 1-a, 2-c, 3-d, 4-a, 5-b, 6-a, 7-b, 8-a, 9-b, 10-a

Short Answers

1. Explain the formation of magnetic domains in a ferromagnetic material.
2. Explain the domain wall movement in the magnetic nanowire.
3. Describe the joule heating effect in magnetic nanowire.
4. Enlist the key features of the racetrack memory.
5. Explain the read/write mechanism in DW-MRAM.

References

1. X. Fong, S. H. Choday, and K. Roy. More than Moore technologies for next generation computer design. Springer, NY USA, 2015. Ch. 6.
2. A. Makarov, V. Sverdlov, and S. Selberherr, "Emerging memory technologies: Trends, challenges, and modeling methods," *Microelect. Reliab.*, vol. 52, no. 4, pp. 628–634, 2012.
3. H.-S. P. Wong and S. Salahuddin, "Memory leads the way to better computing," *Nat. Nanotech.*, vol. 10, pp. 191–194, 2015.
4. H. Li, Y. Chen, Nonvolatile memory design, magnetic, resistive and phase change, CRC Press, NY USA, 2012, Ch. 1.

5. S. Verma, S. Kaundal, and B. K. Kaushik, "Novel $4F^2$ buried-source-line STT MRAM cell with vertical GAA transistor as select device," *IEEE Trans. Nanotech.*, vol. 13, no. 6, pp. 1163–1171, 2014.

6. L. Thomas, S.-H. Yang, K.-S. Ryu, B. Hughes, C. Rettner, D.-S. Wang, C.-H. Tsai, K.-H. Shen, and S. S. P. Parkin, "Racetrack Memory: A high-performance, low-cost, non-volatile memory based on magnetic domain walls," *Elect. Dev. Meet. (IEDM), 2011 IEEE Int.*, pp. 24–24.2.4, 2011.

7. Y. Zhang, W. Zhao, J. O. Klein, D. Ravelsona, and C. Chappert, "Ultra-high density content addressable memory based on current induced domain wall motion in magnetic track," *IEEE Trans. Mag.*, vol. 48, no. 11, pp. 3219–3222, 2012.

8. L. Berger, "Low-field magnetoresistance and domain drag in ferromagnets," *J. Appl. Phys.*, vol. 49, no. 3, pp. 2156–2161, 1978.

9. S. Fukami, M. Yamanouchi, S. Ikeda, and H. Ohno, "Domain wall motion device for nonvolatile memory and logic - size dependence of device properties," *IEEE Trans. Mag.*, vol. 50, no. 11, 2014.

10. Y. Zhang, W. S. Zhao, D. Ravelosona, J. O. Klein, J. V. Kim, and C. Chappert, "Perpendicular-magnetic-anisotropy CoFeB racetrack memory," *J. Appl. Phys.*, vol. 111, no. 9, 2012.

11. S. Parkin and S.-H. Yang, "Memory on the racetrack," *Nat. Nanotech.*, vol. 10, no. 3, pp. 195–198, 2015.

12. G. S. D. Beach, M. Tsoi, and J. L. Erskine, "Current-induced domain wall motion," *J. Magn. Mag. Mater.*, vol. 320, no. 7, pp. 1272–1281, 2008.

13. K. Kawabata, M. Tanizawa, K. Ishikawa, Y. Inoue, M. Inuishi, and T. Nishimura, "Study of current induced magnetic domain wall movement with extremely low energy consumption by micromagnetic simulation," *IEEE 2011 Int. Conf. Simul. Semicond. Proce. Dev.*, pp. 55–58, 2011.

14. S. Fukami, M. Yamanouchi, K. J. Kim, T. Suzuki, N. Sakimura, D. Chiba, S. Ikeda, T. Sugibayashi, N. Kasai, T. Ono, and H. Ohno, "20-Nm magnetic domain wall motion memory with ultralow-power operation," *IEEE Tech. Dig. - Int. Elect. Dev. Meet. IEDM*, pp. 72–75, 2013.

15. G. Tatara, H. Kohno, and J. Shibata, "Microscopic approach to current-driven domain wall dynamics," *Phys. Rep.*, vol. 468, no. 6, pp. 213–301, 2008.

16. W. Zhao, D. Ravelosona, J. O. Klein, and C. Chappert, "Domain wall shift register-based reconfigurable logic," *IEEE Trans. Mag.*, vol. 47, no. 10, pp. 2966–2969, 2011.

17. A. Yamaguchi, S. Nasu, H. Tanigawa, T. Ono, K. Miyake, K. Mibu, and T. Shinjo, "Effect of Joule heating in current-driven domain wall motion," *Appl. Phys. Lett.*, vol. 86, no. 1, pp. 10–13, 2005.

18. K. Huang and R. Zhao, "Magnetic domain-wall racetrack memory-based nonvolatile logic for low-power computing and fast run-time-reconfiguration," *IEEE Trans. VLSI. Sys.*, vol. 24, no. 9, pp. 2861–2872, 2016.

19. H. Nakamura, S. Li, X. Liu, and A. Morisako, "Current-induced domain wall motion in TbFeCo micro wire with perpendicular magnetic anisotropy," *J. Phys. Conf. Ser.*, vol. 266, no. 6, p. 012082, 2011.

20. S. J. Noh, Y. Miyamoto, M. Okuda, N. Hayashi, and Y. Keun Kim, "Effects of notch shape on the magnetic domain wall motion in nanowires with in-plane or perpendicular magnetic anisotropy," *J. Appl. Phys.*, vol. 111, no. 7, pp. 2014–2017, 2012.

21. S. Fukami, T. Suzuki, K. Nagahara, N. Ohshima, Y. Ozaki, S. Saito, R. Nebashi, N. Sakimura, H. Honjo, K. Mori, C. Igarashi, S. Miura, N. Ishiwata, and T. Sugibayashi, "Low-current perpendicular domain wall motion cell for scalable high-speed MRAM," *2009 Symp. VLSI Tech. Dig. Tech. Pap.*, pp. 230–231, 2009.

22. S. S. P. Parkin, "Data in the fast lanes of racetrack memory," *Sci. Am.*, vol. 300, no. 6, pp. 76–81, 2009.

23. S. S. P. Parkin, M. Hayashi, and L. Thomas, "Magnetic domain wall racetrack memory," *Science*, vol. 320, no. 5873, pp. 190–194, 2008.

24. A. J. Annunziata, M. C. Gaidis, L. Thomas, C. W. Chien, C. C. Hung, P. Chevalier, E. J. O'Sullivan, J. P. Hummel, E. A. Joseph, Y. Zhu, T. Topuria, E. Delenia, P. M. Rice, S. S. P. Parkin, and W. J. Gallagher, "Racetrack memory cell array with integrated magnetic tunnel junction readout," *IEEE Tech. Dig. - Int. Elect. Dev.Meet. IEDM*, no. c, pp. 539–542, 2011.
25. H. P. Trinh, W. Zhao, J. O. Klein, Y. Zhang, D. Ravelsona, and C. Chappert, "Magnetic adder based on racetrack memory," *IEEE Trans. Cir. Sys.-I Regu. Pap.*, vol. 60, no. 6, pp. 1469–1477, 2013.

Printed in the United States
By Bookmasters